More praise for *Rewilding Our Hearts*

"In this wise and passionate book, Marc Bekoff brings a lifetime of scientific research and deep personal reflection to bear on our deepening environmental crisis. In his characteristically insightful and engaging style, Bekoff advocates for compassion as the basis of new understandings of ourselves and prompts us to reimagine the kinds of relationships that we might yet have with the rest of our living world. *Rewilding Our Hearts* is a tragically honest and yet powerfully uplifting response to the challenges of our time."

— Thom van Dooren, environmental philosopher,
University of New South Wales, Australia, and author of
Flight Ways: Life and Loss at the Edge of Extinction

"As one has come to expect from Marc Bekoff, this is a wonderful book of scientific stories about animal minds, consciousness, and emotion. But *Rewilding Our Hearts* goes beyond this, showing how wildness in nature, animals, and our hearts are interconnected and mutually reinforcing. This book will bring a new audience to Bekoff's work."

— Dale Jamieson, director of the Animal Studies Initiative and
professor of environmental studies and philosophy, New York University

"Sadly, most of our relations with other animals are dominionistic — transactions between masters and slaves. In *Rewilding Our Hearts* Marc Bekoff argues persuasively that such top-down interactions are bad for creatures, bad for nature, and bad for humans. The key, Bekoff asserts, is to rewild ourselves and *to respect the individuality of other beings and their homes*. When we are unkind to individual creatures, we demean them and hobble our own moral development. Being the most powerful creature, he concludes, does not give us the license to ruin a spectacularly beautiful planet, its wondrous webs of nature, and its magnificent nonhuman residents. Who could disagree with that?"

— Michael Soulé, research professor emeritus in environmental studies,
University of California, Santa Cruz, and coeditor of *Conservation Biology:
An Evolutionary-Ecological Perspective*

"*Rewilding Our Hearts* is about healing — healing our self-destructive attitudes toward the Earth by adopting a more compassionate and ethical treatment of animals. To 'rewild our hearts' is, then, to heal our current dysfunctional and alienated relationship to the planet. Bekoff passionately argues that we must care about other life, and through our caring, we will bring about our own salvation. His argument that we should treat animals with kindness and compassion is an ethical extension of the values that support civil rights — that no humans are truly free until all humans are liberated from repression and fear. In *Rewilding Our Hearts*, Bekoff makes a cogent argument that all of us have both a moral obligation and a need to embrace a new, more encompassing and empathic behavior toward all animals, and that in doing so, we will increase our own well-being and happiness. Read *Rewilding Our Hearts* and you will whoop, smile, and definitely gain a new respect for all life, including your fellow travelers on the planet." — George Wuerthner, ecological projects director, Foundation for Deep Ecology

"Marc Bekoff has been a great pioneer in the scientific study of animal emotion. Now, in *Rewilding Our Hearts*, Bekoff beautifully and simply articulates a philosophical attitude to guide us in the restoration of nature, which has suffered so terribly from human assaults upon it. What's more, Bekoff shows how a new attitude toward nature can help us develop compassion and humility in our own lives. Bekoff's message offers hope for genuine healing, of both our natural environment and ourselves."

— William Crain, professor of psychology, City College of New York, and author of *The Emotional Lives of Animals and Children: Insights from a Farm Sanctuary*

"Marc Bekoff writes about how to harness compassion in a frantic world full of concrete and steel, neon and commerce; a world in which nature is crumbling around us. Finding this book is like finding a pot of glue with which we can fill the hole in our hearts and the ozone." — Ingrid E. Newkirk, president of People for the Ethical Treatment of Animals (PETA)

"Calling for a reawakened caring for nature and animals that is both passionate and informed, *Rewilding Our Hearts* provides the healthy challenge we need in today's critical and confusing times. And Marc Bekoff — mountain man and noted academic — is the right man for this job. A lively, inspiring, and unsettling book that helps us reconnect with our inner wisdom and remember the deeper truths about our shared life on this fragile Earth."

— Will Tuttle, PhD, author of the bestselling book *The World Peace Diet*, recipient of the Courage of Conscience Award, and cofounder of Circle of Compassion

"This book is a reminder that all living things are our family, and we cannot live a spiritually rich and happy life unless we acknowledge and act on our connections. Marc Bekoff suggests compassionate and practical actions, small and large, that we can take to repair our bond with all other species."

— Louise Chawla, professor, environmental design program, University of Colorado, Boulder

"This is a book to make us all think. Drawing as only he can from a wide range of scientific research and personal experience, Marc Bekoff argues that we need to rethink our relationship to animals. *Rewilding Our Hearts* asks humans to give up a little control, act with a little humility, and recognize the connections we have to the rest of nature. Bekoff's examples and illustrations remind us that sharing our planet with other species is a source of resilience as well as delight and that caring for the natural world is essential for human well-being."

— Susan Clayton, PhD, Whitmore-Williams Professor of Psychology and chair of environmental studies, the College of Wooster

"For many years award-winning scientist Marc Bekoff has been writing about the cognitive, emotional, and moral lives of animals. Now, in *Rewilding Our Hearts*, he lays out a practical way for people to reconnect with natural landscapes and animals through understanding and compassion. Far too many of us have become alienated from nature — 'unwilded,' as Bekoff puts it. It is desperately necessary for us to rewild ourselves, to nurture a sense of wonder and a sensitivity to the perspectives and needs of those around us. To this end, Bekoff calls for a rewilding of the education system that would teach young people the importance of connecting with the natural world and stress the need for empathy and compassion for animals and for one another. I believe this is the path toward social justice and peace for all."

— Jane Goodall, PhD, DBE, founder, Jane Goodall Institute, and United Nations Messenger of Peace

"Dr. Marc Bekoff's latest work of art — of compassion in action — is a deeply philosophical, extraordinarily accessible, and powerfully composed message for our times: try to imagine being the person — or dog, rattlesnake, white-tailed deer, prairie dog, horse, tiger, spider, coral reef, rainforest, wolf, fish, wasp, old friend, or 'enemy' — you perceive in order to most effectively empathize with and help her or him. Bekoff reaches back to the most ancient Pythagorean and Semitic concepts of love and translates them into the context of a modern all-out ecological crisis, the worst extinction spasm in tens of millions of years, the most outrageous amount of pain being meted out by humans to hundreds of billions of fellow inhabitants of the Earth. But in doing so, he finds good reason — and plenty of scientific and social-scientific evidence — to be cautiously optimistic, and to rejoice in the prospect of becoming reunited with the millions of other species on this planet. *Rewilding Our Hearts* is a blueprint for compassion that will change the way you think, feel, and act toward others — all others. It is absolutely essential reading for students K–12, as well as undergraduate, graduate, and general readers. A breakthrough work of the highest order. A simply *brilliant* work!"

— Michael Charles Tobias, president, Dancing Star Foundation

"Marc Bekoff is one of the leading compassionate reformers of our time. In *Rewilding Our Hearts*, with his characteristic passionate wisdom, he illuminates a path toward caring connection. This is a book for anyone who is interested in nurturing themselves and healing our alienated relationships with one another, other animals, and the beautiful planet we all call home."

— Lori Gruen, PhD, professor of environmental studies, Wesleyan University, author of "Entangled Empathy," and coeditor of *Ecofeminism: Feminist Intersections with Other Animals and the Earth*

"This inspiring book is impossible to put down. Page after page bristles with stories, examples, challenges; Bekoff offers not only a descriptive ethology, but a poetic sensitivity to resonating and making peace with the natural world that is deeply moving. Above all, he argues that we have lost the deep connectivity to the natural world, and to other animals in particular, that is the basis for a natural ethic, one born of wonder, honed through caring, and developed through wisdom. 'Rewilding,' a term he takes from conservation biology, becomes both a personal and social quest that is much more than simply letting the natural world take its course. Rather, it is about a responsible tuning in to what needs to be done in order to create a less fragile ecological balance, a balance that includes humankind as well. Bekoff's passionate rewilding challenge extends beyond conservation to include global climate change, strategy for city planning, media portrayals of animals, and pedagogy. Above all, he recognizes an implicit spirituality in this process, one that children are naturally drawn to: a humility in the face of the wondrous living earth. Yet this book is not an appeal to romanticism but asks us to make tough choices, knowing the barriers that stand in the way, including misconceived notions such as human exceptionalism. Like other great sages, he offers hope and encouragement to take small steps; it is tiny acts of compassion that count and bring a deep joy to those who live out rewilding."

— Celia Deane-Drummond, professor of theology, University of Notre Dame, Indiana

"*Rewilding Our Hearts* expands our understanding of rewilding beyond the mandate to restore wild nature by enlarging protected areas, interconnecting them, and repatriating their native species. This work is an inspiring call to rekindle our deep-seated connection with the natural world; to awaken our innate capacity for caring about nonhumans and their homes; to participate actively, each of us in the ways we can, in the great work of healing nature and ourselves from our alienation from Earth's living community. Marc Bekoff is a leading voice in the effort to ignite a rewilding social movement grounded in love, compassion, and the recognition that we are inhabitants of a magnificent planet that is calling us home."

— Eileen Crist, coeditor of *Keeping the Wild: Against the Domestication of Earth*

REWILDING OUR HEARTS

Also by Marc Bekoff

Why Dogs Hump and Bees Get Depressed

Jasper's Story: Saving Moon Bears
(a children's book with Jill Robinson)

The Animal Manifesto:
Six Reasons for Expanding Our Compassion Footprint

Wild Justice: The Moral Lives of Animals (with Jessica Pierce)

Animals at Play: Rules of the Game (a children's book)

The Emotional Lives of Animals: A Leading Scientist Explores Animal Joy, Sorrow, and Empathy — and Why They Matter

Animals Matter: A Biologist Explains Why We Should Treat Animals with Compassion and Respect

Animal Passions and Beastly Virtues:
Reflections on Redecorating Nature

The Ten Trusts (with Jane Goodall)

Minding Animals: Awareness, Emotions, and Heart

Nature's Life Lessons: Everyday Truths from Nature (with Jim Carrier)

Edited by Marc Bekoff

The Jane Effect: Celebrating Jane Goodall (with Dale Peterson)

Ignoring Nature No More: The Case for Compassionate Conservation

Encyclopedia of Animal Rights and Animal Welfare

Encyclopedia of Human-Animal Relationships

Encyclopedia of Animal Behavior

The Cognitive Animal:
Empirical and Theoretical Perspectives on Animal Cognition
(with Colin Allen and Gordon Burghardt)

The Smile of a Dolphin: Remarkable Accounts of Animal Emotions

REWILDING OUR HEARTS

Building Pathways of Compassion and Coexistence

MARC BEKOFF

Foreword by Richard Louv

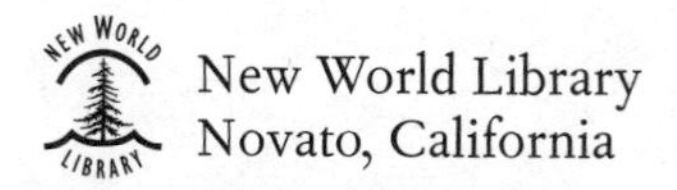

New World Library
Novato, California

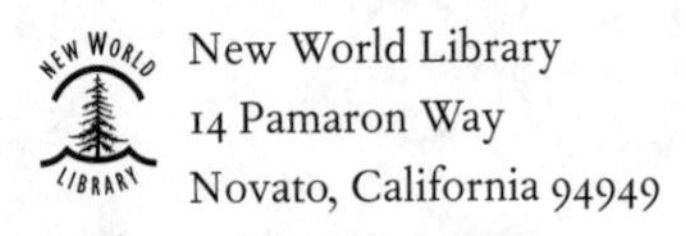
New World Library
14 Pamaron Way
Novato, California 94949

Text design by Tona Pearce Myers
Cover photograph: "Spirit of Denali — Gray Wolf," courtesy of Thomas D. Mangelsen, Images of Nature (www.mangelsen.com)

Library of Congress Cataloging-in-Publication Data is available.

First printing, November 2014
ISBN 978-1-57731-954-2
Printed in the USA on 100% postconsumer-waste recycled paper

New World Library is proud to be a Gold Certified Environmentally Responsible Publisher. Publisher certification awarded by Green Press Initiative. www.greenpressinitiative.org

10 9 8 7 6 5 4 3 2 1

For all animals, nonhuman and human,
may your future be bright with hope, peace, safety, and love
as we rewild and retune our relationship with Earth

Contents

FOREWORD by Richard Louv xiii

INTRODUCTION: What Is "Rewilding Our Hearts"? 1

CHAPTER 1: Global Problems, Personal Unwilding 21

CHAPTER 2: Compassion and Coexistence Mean It's Not All about Us 45

CHAPTER 3: Making It Real: Hard Choices and Bottom Lines 65

CHAPTER 4: Rewilding the Media: Our Mirror Up to Nature 101

CHAPTER 5: Rewilding the Future: Wild Play and Humane Education 119

AFTERWORD: Rewild as You Go 139

ACKNOWLEDGMENTS 151

ENDNOTES 153

BIBLIOGRAPHY 175

INDEX 181

ABOUT THE AUTHOR 197

Foreword

For decades, Marc Bekoff, professor emeritus of ecology and evolutionary biology at the University of Colorado, Boulder, has been at the forefront of our thinking about the relationship between human and nonhuman animals. I first encountered his work in 2000, when I interviewed him for a newspaper column about our evolving cultural attitude about dogs. In a long-term study, his researchers had asked, "What do you think is the biggest problem with open space?" Respondents could choose from among these answers: "too many dogs," "too many people," "too many unruly dogs," "too many unruly people," and "other." More than 90 percent of those surveyed answered, "too many unruly people." He also studied dog and human behavior and found that most of the time when dogs are bad in public, humans have led them astray. His most surprising finding: dogs on a leash tend to bite people more than free-running dogs. "On the leash," he says, "they're more territorial and frustrated." Not unlike people.

In countless papers and books, Marc has illuminated the experiential intersection of humans and other animals. Now, in his remarkable new book, *Rewilding Our Hearts*, he provokes deeper thinking — sometimes comforting, at other times discomforting — about our relationships with other animals and our fellow humans. Not everyone will agree with every position he takes, but one of Marc's distinguishing characteristics is that he can hold strong opinions while remaining open and genuinely curious about the views of others. This admirable quality speaks to the subtitle of this book: *Building Pathways of Compassion and Coexistence*. Here, he stresses the idea of "personal rewilding" as a *compassionate* social movement, one that goes beyond merely connecting with nonhuman animals, to weaving new connections among humans.

I'm flattered to be mentioned in these pages, and I view Marc as one of the leaders of what I've called the New Nature Movement, a movement that includes but goes beyond traditional environmentalism and sustainability; one that maximizes the potential of nature to enhance our minds, our personal and societal vibrancy, and our resilience. This New Nature Movement revives old concepts (in health and urban planning, for instance) and adds new ones. It combines the gifts of both technology and the natural world. It's about saving energy but also producing human energy — in the form of better physical and psychological health, higher mental acuity, and creativity. It's about conservation, but it's also about "creating" nature where we live, work, learn, and play. It's about preserving wilderness but also about transforming cities into engines of biodiversity.

Building such a movement requires more than intellectual excellence, policy debate, or public demonstrations; it demands personal compassion and coexistence with other animals, including other humans. We do not have to agree on everything to move forward to nature. *Rewilding*, as Marc defines it, is, yes, about becoming reenchanted with nature and nurturing our sense of wonder. But it is about more than that. It is about how our connections with other life, human and nonhuman, can change us, make us more compassionate and empathic, and harness "our inborn goodness and optimism." Rewilding is achieved through that most radical of acts: opening ourselves to others. In essence, it is about unleashing our hearts.

— Richard Louv, author of
The Nature Principle and *Last Child in the Woods*

Introduction

What Is "Rewilding Our Hearts"?

When human beings lose their connection to nature, to heaven and earth, then they do not know how to nurture their environment or how to rule their world — which is saying the same thing. Human beings destroy their ecology at the same time that they destroy one another. From that perspective, healing our society goes hand in hand with healing our personal, elemental connection with the phenomenal world.

— Chögyam Trungpa

I've been studying nonhuman animals all my life. As a child, I sensed that other animals had emotions and awareness, and much of my career as a scientist has been devoted to discovering if this was true (it is), and then how and in what ways. Among researchers and scientists today, there is no longer much debate over the fact that many animals are emotional and sentient beings. This paradigm shift has been extremely gratifying to witness, and I'm proud to have played a role in it.

However, over the years, I've also been increasingly concerned with what we do with what we know. How we treat other animals matters a great deal, or at least it should. I have long argued that science and scientists must take into account ethics and morality. It's not enough to say that animals think and feel. This knowledge should affect our actions. It should change how we care for other animals, who, like us, are capable of suffering and also of caring and compassion. And so, I have long made the case, loudly and publicly, that we must give wild and captive (including domestic) animals much more protection than we currently do and recognize and respect them as individuals.

Success in this realm has been much harder to come by. In truth, I believe many individuals would like to see us develop a more compassionate and humane society, one that adequately cares for the thinking and feeling animals in our lives and that preserves and sustains the ecosystems and biodiversity upon which we all depend. But it's not easy to change the entrenched practices of industry, business, science, and society. Not only do individuals have less power to change institutional structures and laws, but people can be reluctant to rethink their everyday lives and habits — to change the clothes they choose to wear and the food they choose to eat, where they shop, and what they do for entertainment — solely to help animals.

Nevertheless, we live in a troubled and wounded world that is in dire need of healing. In fact, the problems we face are so enormous and complex that they dwarf our ability to understand them. They include the ethical obligation to improve the well-being of other animals, but they go far beyond this. Today, the price we face for not changing our ways, and fast, could very well be our own lives. We all should be deeply concerned, perhaps even terrified, by what we have done and continue to do to our planet. We humans — big-brained, big-footed, overproducing, overconsuming, and invasive mammals — have for a long time acted as if we are the only animals who matter. We have made huge and horrific global messes, impacting every environment and ecosystem and all other species, and these effects need to be halted and, if possible, repaired. No one truly knows how many of the changes we've wrought are irreversible. But if we don't change, we will certainly only continue along our current self-destructive path.

This is the issue I want to address in *Rewilding Our Hearts*. In many ways, this book is a natural successor to some of my previous books, including *Minding Animals*, *The Emotional Lives of Animals*, *The Animal Manifesto*, and *The Ten Trusts*, which I cowrote with renowned conservationist and primatologist Jane Goodall. In these, I also focused on the well-being of nonhuman animals (as distinct from humans, who are also animals, though in this book I sometimes use just the word "animals" for convenience). However, in significant ways, this book is also different than any I've written before. Here I am more focused than ever on us: on how we think and how we need to change our thinking. In *Rewilding Our Hearts*, I look at the big picture of the dilemmas facing our age — namely, climate change, our population explosion, and the consequent damage to Earth's ecosystems and loss of biodiversity — and suggest the mind-set that we need to adopt in order to fix them.

This book is also different because more than ever I'm speaking from my heart as a concerned global citizen, not primarily as a scientist, though science plays a large role in the process of rewilding. *Rewilding Our Hearts* calls for another paradigm shift, one that values compassion above all. My hope is that, if we as a species can proceed from compassion, such that the life of every *individual* animal matters whenever we make choices, societal or personal, we might begin to undo the alienation and fragmentation that currently defines our damaged relationship to the natural world. Compassion will also help us heal our damaged, alienated, and fragmented relationships with each other. I believe that only then, when we hone peaceful coexistence for all beings, human and nonhuman, will we be able to find the ecological solutions that we so desperately need.

Rewilding Begins with Caring

Our world is becoming global. Many of us are in daily contact with people from other parts of the world, other cultures, and language groups. As globalization proceeds, we will need, urgently, every ounce of empathy, tolerance, and communication skill we can muster. This is what our connection with animals has given us; this is what we need so badly for the future.

— Pat Shipman

It is impossible, or should be impossible, to be neutral on issues of animal well-being, environmental protection, and the effects of climate change. Many scientists like to think science is objective and that they themselves don't have an agenda. Many also think they have no obligation to interact with nonresearchers, that is, the public. But even if the aim of science is objective discovery, scientists are subjective. They are thinking, feeling humans with particular viewpoints and a stake in the world, which is how it should be. In his wonderful and bold book *A World of Wounds*, renowned biologist Paul Ehrlich wrote: "Many of the students who have crossed my path in the last decade or so have wanted to do much, much more. They were drawn to ecology because they were brought up in a 'world of wounds,' and want to help heal it. But the current structure of ecology tends to dissuade them."

The concept of rewilding is grounded in the premise that caring is okay. In fact, it is more than okay; it is essential. It is all right to imagine the perspective of nonhuman animals in order to take their well-being into account. People who care about animals and nature are often made to feel they must apologize for their views. They are disparaged for "romanticizing" animals or being sentimental, and they can be presented as "the radicals" or "the bad guys" who are trying to impede "human progress." This book takes a hard look at that progress, and it proposes that by fighting for and caring for animals better, we will also be fighting for and caring for humanity better. When I was in Sydney, Australia, in February 2013, I had a stimulating talk with environmental scientist and writer Haydn Washington. After I mentioned that I was sick and tired of people who cared about animals having to apologize for their compassion and empathy, he said something that really moved me: "People should not have to apologize for their sense of wonder." Amen.

Rewilding our hearts is about becoming reenchanted with nature. It is about nurturing our sense of wonder. Rewilding is about being nice, kind, compassionate, empathic, and harnessing our inborn goodness and optimism. In the most basic sense, "rewilding" means "to make wilder" or "to make wild once again." This means many things, as we will see, but primarily it means opening our hearts and minds to others. It means thinking of others and allowing their needs and perspectives to influence our own.

A long time ago I developed the notion of "minding animals." I still like this phrase and use it in two main ways. First, "minding animals" refers to caring for nonhuman animals, respecting them for who they are, appreciating their own world views, and wondering what and how they are feeling and why. We mind animals when we try as hard as we can to imagine their point of view. But the phrase also acknowledges and honors the well-established fact that many animals have very active and thoughtful minds. We may never know everything that nonhuman animals think and feel, but they see and react to the world with awareness and emotion. In the same way, we can also "mind Earth." We must care for her and appreciate, respect, protect, and love her. To do this, we must imagine the Earth's perspective, which is to say, the collective perspective and well-being of all her inhabitants.

In the end, we are "all one." All beings and all landscapes connect and interact in reciprocal ways. This is a social and ecological truth. What happens in my hometown of Boulder, Colorado, influences and is influenced by what happens outside of Chengdu, China, where I go to work in Animal Asia's Moon Bear Rescue Centre. Minding animals and minding Earth in this way increases our wisdom. It helps us not by accumulating more objective "facts," but by helping us make wiser choices about our actions in the world. Wisdom recognizes that our actions always impact others, even others we don't currently recognize or see, that we always make choices in how to live, and that there are better and worse decisions, which have vastly different impacts. As anthropologist Pat Shipman says, it is imperative that we make decisions today with global impacts in mind.

When I mind animals in this way, I practice what I consider "deep ethology." That is, as the "seer," I try to become the "seen." When I watch coyotes, I become coyote. When I watch penguins, I become penguin. I will also try to become tree and even rock. I name my animal friends and try to step into their worlds to discover what it might be like to be a given individual — how they sense their surroundings, how they move about, and how they behave in myriad situations. This isn't just a flight of fancy. These intuitions can sometimes be the fodder for further scientific research and lead to verifiable information, to knowledge. As a scientist, I know that it's never enough to simply imagine another animal's perspective. But as a person, I know that it's never enough to accept unclarity or uncertainty about animal minds as a reason not to care for them, or as an excuse for inaction or willful harm. This distinction is important for this book.

Rewilding Our Hearts is also more interdisciplinary than my previous books. As I consider a more global and holistic perspective, I go beyond my specific areas of expertise in ethology and compassionate conservation to consider biology, psychology, sociology, philosophy, and anthropology. All of these, in fact, inform studies of animal behavior, animal minds, and conservation biology, but they are not always acknowledged. I also speak as an activist arguing for social change. Ultimately, what we need along with more information about animals is a social movement and revolution in how we interact with animals and nature, a movement based on peace, compassion, empathy, and social justice. As you will see, my vision of this movement is not that it represents a single idea or a specific program. There is no "membership." Instead, we are all already members, as living, breathing human beings who move in circles of coexistence. Peace, compassion, empathy, and social justice are all part of a much-needed revolution in thinking and acting with kindness for all.

Rewilding Our Hearts is a positive and inspirational book about what we can and must do, as individuals within a global community, working in harmony for common goals, to deal with the rampant and wanton destruction of our planet and its innumerable and awe-inspiring residents and their homes. We really do need wild(er) minds and wild(er) hearts to make the changes that need to be made right now, so that we can work toward having a wild(er) planet. The Earth is tired and broken and isn't infinitely resilient. Like a fatigued person who is teetering on burning out, our wondrous and magnificent planet needs all the help it can get. I see wild as beautiful, but as Terry Tempest Williams writes so eloquently, today we must often attempt to find "beauty in a broken world."

Rewilding is a transformative and a personal process. It is a call to action, but primarily to action within our own lives. It is a lens, a way to view the world, that suggests that the combined strength of our individual personal journeys can harness a new global social movement that will help all beings, human and nonhuman alike.

The Roots of Rewilding: Ecological Repair

Rewilding is all the rage in conservation circles.

— *New Scientist*, March 1, 2014

The word "rewilding" became an essential part of the dialogue among conservationists in the late 1990s, when two well-known conservation biologists, Michael Soulé and Reed Noss, wrote a now-classic paper called "Rewilding and Biodiversity: Complementary Goals for Continental Conservation." In her book *Rewilding the World*, conservationist Caroline Fraser noted that rewilding basically could be boiled down to three words: cores, corridors, and carnivores. That is, according to Dave Foreman — the director of the Rewilding Institute in Albuquerque, New Mexico, and a true and some might say radical and occasionally irreverent visionary — rewilding is a conservation strategy based on three premises: "(1) healthy ecosystems need large carnivores, (2) large carnivores need big, wild roadless areas, and (3) most roadless areas are small and thus need to be linked." Conservation biologists see rewilding as a large-scale process involving multiple projects of different sizes that may focus on carnivores but ultimately include the panoply of wildlife.

One such ambitious, courageous, and forward-looking effort is the Yellowstone to Yukon Conservation Initiative, known as the Y2Y project. This science-based project is all about connectivity. It involves maintaining and building natural corridors — replete with underpasses and overpasses around roads and creating protected areas near human communities — so that animals can move about freely and safely without having to worry about people or cars. The Y2Y project also is concerned with preserving native plants and allowing human communities to thrive. Thus, its goals are inclusive of all beings, nonhuman and human. As clearly stated on their website, "Y2Y works to promote best practices for use of the landscape so wildlife and humans can successfully coexist in this region together."

Obviously, ecological rewilding efforts center on the difficult question "What is wild?" Our notions of "wild" and "wildness" are moving entities. In modern times, what people consider to be wild may be considerably different than what earlier humans took to be wild or what future humans will consider to be wild. "Wild" can mean unchanged, untamed, unexplored, and unfamiliar, but in ecological terms, few places on Earth still qualify. Wild can also mean uncontrolled or out of control. But by definition, conservation biologists deliberately alter an environment through rewilding; they don't simply let whatever happens happen in a landscape. Most particularly, the rewilding process must include control or management of ourselves because we humans are all over the place, and we have radically changed natural and evolved checks and balances in the behavior of many different animals and ecosystems across the globe (for example, see *Behavioural Responses to a Changing World*, edited by Ulrika Candolin and Bob Wong).

Further, reintroducing wildlife — rebuilding the biodiversity of animal and plant life in an ecosystem — will never fully or exactly replicate what an environment was like before. So, contrary to what the term implies, rewilding projects can't, and don't, re-create what was before in a manner that is unmanaged by humans. Strictly speaking, then, these ecological efforts don't "rewild" to achieve what was. But what they do is to make room for much more diverse, healthy, and sustainable ecosystems that are as natural as they can be given our omnipresence. They account for the needs of a wider range of creatures, and they adjust human impacts so that this wilderness community can thrive even though it remains constricted by and adjacent to human communities.

Rewilding projects necessarily must involve many different people, who must work together as a community to realize the conservation effort. For large-scale rewilding projects, success always comes down to the core word "corridor." Projects must establish connections or links among diverse geographical areas so that animals can roam as freely as possible with few if any disruptions to their movements. Ecosystems must be connected so that their integrity and wholeness are maintained or reestablished. Regardless of scale, reconnecting a wider variety of habitats, and protecting these corridors and core spaces, helps ecosystems heal themselves and find a new, healthier, less-fragile balance. In other words, this connectivity fixes the central problem, that nature has become too fragmented, and the areas that remain protected (or are not yet exploited) are too isolated from one another.

Interestingly, when rewilding projects create these corridors on behalf of nature, they inevitably reconnect diverse and sometimes fragmented human communities as well. We wake up to the fact that our roads, bridges, businesses, and towns influence and impact our neighbors, and together we must balance the relative costs of these impacts for all stakeholders, including nature.

When I was a graduate student at Washington University in St. Louis, Missouri, I had the distinct pleasure of talking with, and especially listening to, Barry Commoner, one of the founders of modern ecology. Often called the "Paul Revere of Ecology," Barry was far ahead of his time in his broad view of ecology. Most of all, he saw how issues of social justice and concerns for the environment were intimately connected, and I still remember going home after our chats with my mind on fire with new ideas. Barry's four rules of ecology were: "Everything is connected to everything else. Everything must go somewhere. Nature knows best. There is no such thing as a free lunch."

These ideas are central to rewilding today, particularly the ideas that everything is connected and every choice entails a cost.

Rewilding as Life Strategy and Personal Journey

A human being is a part of the whole, called by us "Universe," a part limited in time and space. He experiences himself, his thoughts and feelings as something separated from the rest — a kind of optical delusion of his consciousness. This delusion is a kind of prison for us, restricting us to our personal desires and to affection for a few persons nearest to us. Our task must be to free ourselves from this prison by widening our circle of compassion to embrace all living creatures and the whole of nature in its beauty. Nobody is able to achieve this completely, but the striving for such achievement is in itself a part of the liberation and a foundation for inner security.

— ALBERT EINSTEIN

By proposing that we "rewild our hearts," I mean just that. Rewilding personalizes what conservation projects try to accomplish in the world by building wildlife bridges and underpasses so that animals can move freely between fragmented areas. I see rewilding our hearts as a dynamic, intimate process that fosters corridors of coexistence and compassion for animals and their homes at the same time that it facilitates corridors in ourselves that connect our heart and brain, our caring and awareness. In turn, these connections, or reconnections, will result in making wiser choices, pursuing heartfelt actions that make the lives of all beings better. Rewilding our hearts and rewilding the human dimension mean redefining the borders in our interactions with other animals and overcoming the cognitive dissonance that abounds globally.

I like to use the word "borders" rather than "boundaries" or "barriers" because the latter words imply a less permeable interface between "them" and "us." Redefining and softening these borders and distinctions is what rewilding is all about. Because of this, rewilding demands that we employ humility in our interactions with other animals and their homes. We really are the dominant and dominating species, and so, to achieve more equality in our interactions with nature, we need to control ourselves. We need to "be humble in the face of nature's awesomeness." We should regard and approach nature as a friend, one whose welfare matters for its own sake — and even more so because it matters for our sake, too.

Thus, "rewilding" is a mind-set. It reflects the desire to (re)connect intimately with all animals and landscapes in ways that dissolve borders. Rewilding means appreciating, respecting, and accepting other beings and landscapes for who or what they are, not for who or what we want them to be. It means rejoicing in the personal connections we establish and need so badly. Indeed, I see the process of rewilding as, most of all, a personal journey and transformative exploration that centers on bringing other animals and their homes, all ecosystems, back into our heart. It is inarguable that if we are going to make the world a better place now and for future generations, personal rewilding is central to the process. Laws and public policy won't do it. Instead, each of us must undergo a major personal paradigm shift in how we view and live in the world and how we behave. Researchers agree. For example, conservation biologists Andrew Balmford and Richard Cowling stress that "conservation is primarily not about biology but about people and the choices they make." California State University psychologist

P. Wesley Schultz notes, "Conservation can *only* be achieved by changing behavior."

In chapters 1 and 2, I discuss further what this new mind-set entails, and in chapters 3 to 5, I look at where we most need to change — in our society, culture, and education — and how re-wilding our hearts can make a difference. There is no single solution to our many and various problems; in fact, there is rarely a single solution to any one problem. We need to be flexible and open at all times. In this way, rewilding is like a moral compass we can use as we chart our collective path into a challenging future. On this journey, we face many vexing questions that we can't avoid and perhaps don't even want to address, but re-wilding encourages us to meet the world honestly and directly and to move out of our comfort zones and think out of the box. For instance, wild also means "unpredictable," and anything that is unpredictable, any outcome we can't guarantee or accident from which we can't protect ourselves, makes us nervous. So, to rewild ourselves means to accept a certain level of risk and uncertainty.

Rewilding helps us be comfortable with our occasional twinges of fear and anxiety and not project them onto others. When I lived in the mountains outside Boulder, Colorado, I would sometimes feel this fear when walking or riding along the roads and trails around my house. Some animals can be dangerous; I might unexpectedly startle them, or they might not appreciate my presence in their home. I had to be on the alert. This isn't a bad thing, and it always reminded me how incredibly fortunate I was to be sharing my home with other magnificent animals. I chose to live where I did, and if I didn't like the particular terms of coexistence, I could move elsewhere.

In fact, the good feelings that come from rewilding far outweigh the anxieties and costs. This is another way that rewilding can be seen as a moral guide. When I see my actions as being helpful to others, a warm feeling engulfs me. I feel good to know that I am doing what I can to coexist with animals and to improve the welfare of the environment, our shared home. To want to help others in need is natural, and experiencing that positive glow is an important aspect of rewilding.

The Planet of No Return

Conservation is humankind's attempt to protect nature from ourselves, yet we are not separate from it. We often forget that we're a part of nature in an interdependent albeit sometimes adversarial relationship. Even a New York City girl needs a connection to the wild.

— Mark Hoelterhoff

Over the past few years, I have come to see that we are always "re-ing" one thing or another. Rewilding in the real world requires us to try to restore and re-create ecosystems, for example, by reintroducing or repatriating animals into areas where they once lived. But since we really cannot re-create or restore ecosystems "to what they were," it has been suggested that we rebuild rather than rewind as we move into the future. Rebuilding surely is part of the process of rewilding. We also talk about the need to rekindle, rebalance, refine, reconnect, reenvision, reintegrate, reimmerse, reeducate, rehabilitate, rethink, and reshape our relationships with other nature (see, for example, the collection of interdisciplinary essays in *The Politics of Species*, edited by Raymond Corbey and Annette Lanjouw). Many of these efforts are reactions to environmental and ecological problems we can no longer ignore, but being reactive — the "putting out the fire" mentality — does not work. As we rewild our hearts, there is an urgent need to be proactive. Instead of looking to the past as a guide, we have to envision the positive future we want and actively work toward it.

Erle Ellis, who works in the Department of Geography and Environmental Systems at the University of Maryland, suggests that we've transformed Earth beyond recovery. He says Earth has become "the planet of no return." Renowned environmentalist Bill McKibben says much the same thing: "We're not... going to get back the planet we used to have." In other words, whatever happens next, the world will never be the way it was. Ellis writes:

> There will be no returning to our comfortable cradle. The global patterns of the Holocene have receded and their return is no longer possible, sustainable, or even desirable. It is no longer Mother Nature who will care for us, but us who must care for her. This raises an important but often neglected question: can we create a good Anthropocene? In the future will we be able to look back with pride?... Clearly it is possible to look at all we have created and see only what we have destroyed. But that, in my view, would be our mistake. We most certainly can create a better Anthropocene. We have really only just begun, and our knowledge and power have never been greater. We will need to work together with each other and the planet in novel ways. The first step will be in our own minds. The Holocene is gone. In the Anthropocene we are the

creators, engineers and permanent global stewards of a sustainable human nature.

In the 1980s, scientists suggested renaming our current geological epoch the "anthropocene," or roughly, the age of humanity, as a way to acknowledge the epochal changes the human species has enacted on the Earth. The last dozen millennia have been marked by overriding turmoil, which is apparent to anyone who takes the time to pay attention. Researchers and nonresearchers alike are extremely concerned about unprecedented global losses of biodiversity and how all creatures, including humans, and a myriad of landscapes are suffering because of our destructive ways. We are animals — we are great apes — and we should be proud of our membership in the animal kingdom. However, our unique contribution to the wanton decimation of the planet and its many life forms is an insult to the Earth that also demeans us as a species. Of course, we should care about this very much.

In other words, though rewilding our hearts is an intimately personal process, its aims are broader and deeper than our personal lives. We must work on long-term solutions of different scales that encompass our society and the globe, for the Earth is really our one and only home. We need to think out of the box and become "uncommon messengers," since what we've been doing up to now has not worked very well. Today, with seven billion of us and counting, our problems will only get bigger and more difficult to solve. If we continue to ignore, fragment, and fracture nature, we will continue to do the same to ourselves. In our denial, we may not even realize we're self-destructing, but we need nature just as we need oxygen.

To achieve this wider perspective, we need what Ian Billick

and Mary Price call an "ecology of place" (in their book of the same name), which focuses on producing local as well as general knowledge. This perspective stresses an interdisciplinary approach and seeing the world as an interconnected community. In a way, it takes literally the well-known activist creed "think globally, act locally." We must take local actions, not only on faith that they will help the world, but in ways that take the world into account and ideally have widespread impact. Our very existence depends on our paying close attention to *all* of nature and her intricate tapestry of diverse and magnificent landscapes, which is the exact opposite of what happens when we ignore and compartmentalize nature and exempt ourselves from being members of a vast and interconnected community. Simply put, when we feel disconnected from our responsibility as a member of sublime and grand nature, we become alienated and we suffer.

One of my favorite bumper stickers is "Nature bats last." We can try to outrun and outsmart nature, but in the end she always wins. We know full well that species come and go, and nature survives. She evolves and moves on. Will we let ourselves become one of the species who didn't make it? Or worse, will we continue as the one species who threatens all others, and who lets uncounted species perish, in order to live where and how we please? I hope not. And I trust that this book makes a clear and strong case that we need a heartfelt revolution in how we think, what we do with what we know, and how we act. The revolution has to come from deep within us and begin at home, where we live. Despite our diversity and our differences, if we rewild our hearts with empathy and compassion for all nature, we will find common ground, effective compromises, and practical solutions.

Who could be against a global social movement based on peace, compassion, empathy, and love? Gentle kindness and humility can lead us out of the murky waters in which we are now trying to survive and thrive. We need to establish tight networks of compassion that cross cultures. We simply cannot leave things to chance. We are an exceptional and unique species, and so too are all others.

Despite all the rampant destruction that is going on globally, there are many reasons for hope and to keep our dreams alive. Numerous people around the world — and I meet many of them in my travels — are working hard to make the world a better place for all beings. There really is hope, particularly if we can surrender to the deep-rooted feelings of compassion and empathy that emerge when we connect and reconnect with nature, when we rewild our hearts. We really do feel good when we do nice things for others — when we accommodate their needs and welfare. By rewilding, we get past the false dualisms of us and them, humans and animals, civilization and nature. Rather, we see and cultivate a close and reciprocal interrelationship built on peace, compassion, empathy, and love. There is only all of us sharing a single home that supports everyone. If we allow it to, rewilding our hearts comes naturally — since it is who we and other animals truly are.

Because I am thoroughly obsessed with human-animal interrelationships and the fields of animal behavior, cognitive ethology (the study of animal minds), conservation biology, and compassionate conservation, I thought this book would be easy to write. But I was so wrong. It's always been that the most difficult essays and books to write are those that are closest to my heart. Other authors have told me that this is rather common.

I have been thinking about human-animal relationships for decades, and the issues are wide-ranging and require a good deal of deep and broad thinking and discussion. Nonetheless, one thing is very clear: Other animals deserve to live safe and peaceful lives and to thrive because they are alive, because of *who* they are, not because of what we want them to be.

1. Global Problems, Personal Unwilding

Bereft of contact with wildness, the human mind loses its coherence, and the human heart ceases to beat.

— David Abram

We have conquered the biosphere and laid waste to it like no other species in the history of life. We are unique in what we have wrought.

— Edward O. Wilson

Humans are an unprecedented force in nature. We are all over the place, and there are far too many of us. There is hardly anywhere on Earth, in the water, or in the sky that has not been influenced by us. No need to look for mythical Bigfoot: We're here! We leave huge footprints wherever we go, which create all sorts of urgent global problems that so far we have been rather unsuccessful at solving.

This chapter looks at some of these problems and, just as important, their emotional effect on us. Much has already been written about the messes we make, and so I will only summarize them briefly, but it's important to describe what needs fixing. However, the main message is simply put: It is essential that we stop ignoring nature. Animals aren't "ghosts in our machine," invisible objects with whom we can do whatever we choose. And landscapes aren't infinitely resistant and resilient. We must pay close attention to what we are doing and to the incredibly wide-ranging influence we have on our planet. Geologically speaking, the human species is just a blip in time, but there is no doubt that we are the most influential species, in good and bad ways, that has ever existed. Changes to our planet for which we are directly responsible are happening more rapidly and are more widespread than ever before.

Biologist Robert Berry fears we're simply "running out of world." Others argue that we have created a world that is so technologically and socially complex we simply cannot control it, while others claim that in our rapidly changing world concepts such as "natural" make little sense. Perhaps the same can be said about wild, wildness, and wilderness. Certainly, if we define "wilderness" by what the Earth was like even two or three thousand years ago, we'll never see that again so long as humanity survives, and it does us no good to fantasize about what the good old "wild" days were like. What we consider "wild nature" today is largely artificial.

Then again, one suggestion has been made to create a "world park" using "the last remaining 12 percent of healthy natural areas, wilderness areas, primary ecosystems, mini parks, and hotspots." This is the sort of idea that could be part of a global rewilding strategy. It is important to keep the wild as wild as it can be (see *Keeping the Wild*, edited by George Wuerthner, Eileen Crist, and Tom Butler).

At a symposium on biodiversity, conservation, and animal rights held in March 2012 at the School of Oriental and African Studies at the University of London, ecologist and award-winning filmmaker Michael Tobias referred to the places on Earth where we have had the most devastating effects as "pain points." He noted that there are many "pillars of pain" on Earth, some right in our own backyards. It can be very hard to acknowledge this pain and accept our responsibility for it, which leads to the alienation and denial that often undermine efforts to fix these places. Yet our omnipresence and power call for humility and responsibility; this is the attitude of rewilding. Thankfully, connecting with the natural world, and caring for it, comes naturally and feels good. It heals the Earth and us from the inside out.

Overpopulation and Overconsumption

Environmental problems have contributed to numerous collapses of civilizations in the past. Now, for the first time, a global collapse appears likely. Overpopulation, overconsumption by the rich and poor choices of technologies are major drivers; dramatic cultural change provides the main hope of averting calamity.... Humankind finds itself engaged in what Prince Charles described as "an act of suicide on a grand scale."

— Paul and Anne Ehrlich

While overconsumption has been a hot topic for years, many people remain hesitant to address overpopulation. But the plain fact is we are making too many babies. Until the human species stops growing, it will be virtually impossible to cut back on our overall consumption of the Earth's resources. Both issues go hand in hand. Talking about and doing something to curb the rapid rise in human numbers is an essential part of the process of rewilding the world.

Overpopulation is the perfect example of a thorny global issue that is overwhelming for individuals to consider and that defies our ability to develop a coordinated response that everyone will like. Nevertheless, it is occasionally being addressed head on, such as in the book *Life on the Brink: Environmentalists Confront Overpopulation*, edited by Philip Cafaro and Eileen Crist. We need to face the fact that there are too many of us and that we must do something about it right now. The world population is currently over seven billion people, and by 2050, we should reach about nine billion. So, over the next thirty to forty years, that's two billion more mouths to feed and people to house. Where will they live, and how much will they consume? Where will we find the energy, timber, clothing, food, and space? It's also well-documented that as there are more of us there are far fewer of "them," that is, other animals. Overpopulation is a key factor in species extinction (see below), so to solve the latter, we must solve the former.

Our supersized brains should caution us that we cannot go on living as we have, but something doesn't seem to click. For one thing, it is not just sheer numbers that should concern us, but the exponentially faster rate at which we are multiplying. As a species, we are expanding too fast to keep up with ourselves and establish a sustainable equilibrium. I like how Warren Hern, a local Boulder physician, puts it:

> The human population continues to grow and grow and grow and grow.... We have added the most recent billion

> people to the human population in less than 12 years, [and] the human population has doubled in the past 44 years (or less). But in prehistory, it took 100,000 years or more to double. At this rate, we will reach about 13 billion by 2050 and 25 billion by the end of this century. Most population experts dismiss this possibility, but population experts in 1925 said that the human population would never reach 2.5 billion. We passed that number in 1949. My mother, who is 94, and her sister, who is 97, have seen the world population quadruple in their lifetimes. I was born in 1938, and it has tripled in my lifetime. Before now, no human being ever saw that happen. This is a unique time in human evolutionary history.

So, it's essential that we take lessons from nature and other animals and find a way to manage our own population so that we live within our means. If we don't regulate our population size proactively on our own, nature will eventually and surely do it for us. This is the lesson of other species that have overrun their environments. As a species, our current reproductive strategy is really insane and unsustainable and clearly spells doom for us, and it's a prime cause of the rampant collateral damage to the Earth and other creatures. To get us to think about overpopulation and its ecological effects, the Center for Biological Diversity is now giving out condoms in colorful packages depicting endangered animals.

These are the sorts of connections that rewilding makes. Using birth control is not just a matter of practicing safe sex; it can be seen as one very personal way to save the environment. In itself, increased worldwide access to birth control for everyone would help tremendously, as it would curb unwanted pregnancies, but it's equally important just to draw the link between having babies and environmental impacts, climate change, and species extinction. If smaller families were presented as an ecological good and even a necessity for our ultimate survival, more people might choose to have fewer or no babies. Few people want or enjoy family limits that are imposed on them by government, as China has done by enforcing one-child families. To succeed, any request of personal sacrifice needs to be seen as fulfilling an undeniable, agreed-upon social good. This is the challenge of rewilding, and perhaps of the future of our species. It's why we must make personal rewilding all the rage.

Michael Soulé, founder of the field of conservation biology, perhaps captured our population predicament the best: "We're certainly a dominant species, but that's not the same as a keystone species. A keystone species is one that, when you remove it, the diversity collapses; we're a species that when you add us, the diversity collapses. We can change everything, dictate everything and destroy everything."

Climate Change: The Ice Is Melting

Ecosystems are not only more complex than we think,
but more complex than we can think.

— Frank Egler

Climate change, a.k.a. global warming, is a huge topic, one that is constantly in the news, and yet there are strong indications that we have seriously underestimated just how bad it really is. In November 2012, scientists in my hometown of Boulder, Colorado, working at the National Snow and Ice Data Center discovered that polar ice is melting at a record high. For example, polar ice sheets are melting three times faster than in the 1990s. This means Arctic ice has hit a record low, and this carries numerous disastrous environmental implications, from rising seas to a wide range of ecological and wildlife impacts.

In *The Ten Trusts*, Jane Goodall relates a conversation with Angaangaq Lyberth, the then-leader of the Eskimo nation from Greenland. At a gathering of a thousand religious and spiritual leaders in the United Nations General Assembly Hall, he said: "In the north, we feel every day what you do down here. In the north the ice is melting. What will it take to melt the ice in the human heart?"

Rewilding is about melting the ice in our hearts so that we might all work together to solve the dilemmas posed by climate change. First, rewilding asks us to recognize the connection between what we do and its effect on the Earth's changing climate. Global warming is, of course, a very complex collection of interrelated impacts, one that is almost beyond our capacity to understand in its entirety, but we can no longer deny that it is happening and that humans influence it. Those who still deny human-caused climate change are in the same camp with the very few remaining skeptics who argue we really do not know if other animals are conscious or emotional beings. Given the wealth of scientific data on both issues, this skepticism — or as some might call it, agnosticism — is antiscience and harmful to animals and to us. This attitude is "problematizing the unproblematic," as Indiana University philosopher Colin Allen, my esteemed colleague and cycling buddy, puts it about the issue of animal consciousness. At the very least, as I've said many times, the precautionary principle should lead us to deal proactively with the issue of climate change, rather than wait till we're 100 percent certain of all the causes and find it's too late to act.

Indeed, individuals of innumerable species are struggling to adapt to the changes that humans make. Climate change may already be beyond our ability to stop or manage it, and the best we can do is to halt the activities that make it worse and find ways to adapt to our changing environment. This means, of course, doing what we can to help other animals as they are impacted by changing climate. As I make clear in the next section, our success as a species depends on the Earth maintaining a sustainable biodiversity, and so it's in our interest to care about whether other animals can adapt. Some species, perhaps many, will not. Already, American pikas and polar bears have become symbols of the devastating effects a warming climate can have on the lives of animals. It affects movement patterns, social behavior (including mating), social organization, food availability, and interactions among difference species.

Specific examples of these impacts accumulate almost daily. For instance, we know that hot weather lowers the survival of Asian elephants, and that male painted turtles are imperiled by warming temperatures. Lobsters in Maine are also larger, and there are more of them, because of climate change. The last decade was the warmest on record in the Gulf of Maine, and so the lobsters are eating one another and causing a decline in the profits for lobstermen. Starfish sacrifice an arm in order to survive in warm water, and rising seas will wipe out resting and refueling sites for migratory birds. In fact, approximately half of living bird species are threatened by climate change.

We're just beginning to unravel how climate change influences the behavior of other animals. Two discoveries serve notice that the effects can be subtle, unpredictable, and extremely important. In one, it's been discovered that ocean acidification can reverse the response of nerve cells of Australian damselfish, so that the scary and aversive scent of predators suddenly becomes alluring, leading damselfish to become increasingly bold. This behavioral change — a fish approaching rather than avoiding dangerous predators — certainly can't be good. In another example, it's been found that three-lined skinks, a type of lizard, become superintelligent when they develop in warmer temperatures because of changes in how the brain develops. This might be seen as good for three-lined skinks, but it's a sobering indication of how little we understand about how environment affects who animals are, much less an environment that is rapidly changing before our eyes.

Thus, rewilding means we should take a "Noah's ark" attitude toward climate change. As seas rise and the environment changes, we should each do what we can, as we can, to preserve, protect, and conserve all species, since we are all in the same boat, and we need one another.

Where the Wild Things Were: Loss of Biodiversity

Animals are vanishing before our eyes, and people all over the world are asking, "Where have all the animals gone?" They miss hearing, seeing, and perhaps smelling them. This issue always makes me think about Rachel Carson's wonderful book *Silent Spring*, and also Will Stolzenburg's book *Where the Wild Things Were*. Species extinctions are occurring at a dangerous rate, and every year more species become threatened or endangered. I consider the loss of wild animals as a form of abuse, but it rarely is understood or considered in this light. Animal losses not only put the Earth and her ecosystems in peril but they are detrimental to our own well-being and survival.

Biodiversity is what enables human life. This is such an accepted ecological fact that it doesn't need further proof. Thus, it is imperative that all of humanity reconnects with other animals and fights for the survival of every species, for all species need one another. As an interdependent species, this is nothing less than a collective fight for our own survival. When animals die, we die, too. In this way, rewilding contains an element of selfishness: By making the world a better place for all, we are helping ourselves.

To put the larger issue of species extinctions in perspective, consider these details. According to the Center for Biological Diversity, we're experiencing the worst loss of species since the dinosaurs died off 65 million years ago. We're losing species at around a thousand to ten thousand times what's called the "background rate," or the rate of extinctions that would be expected to occur naturally. Amphibians are the most endangered animals. Their current rate of extinction has been estimated to range from twenty-five to forty-five thousand times the background extinction rate. This loss influences the many habitats in which amphibians live because of changes in their consumption habits, their nutritional cycling, and their role in controlling pests, among other things. For instance, as amphibians are lost, humans might suffer from an increase in harmful insects.

In addition, the "biodiversity boom" in Madagascar — the formation of new species — has slowed, and humans cause over half of the death rates (about 52 percent) in North American populations of large and medium-sized mammals. We cause the most deaths (over 34 percent) in larger North American mammals, including those living in protected areas. There also has been a dramatic decline in suitable habitat for African great apes.

That said, we really have no idea how serious the issue of species extinctions really is because we have no idea how many species there currently are. For example, a new species of bird, called the Cambodian tailorbird, was recently discovered in Phnom Penh, the capital of Cambodia. Who would have thought we could overlook a species living within a city? However, life in a fast-moving city can make us unaware of the presence of other animals, not to mention other people. In 2012, among the new animal species discovered were a butterfly in Jamaica, a tarantula in Brazil, a skink in Australia, the Lesula monkey in Democratic Republic of Congo, and a meat-eating sponge in California's Monterey Bay. One source claims that an astonishing 86 percent of all plants and animals on land and 91 percent of those in the seas have yet to be named and cataloged. Sadly, this revelation only underscores the fact that, in all likelihood, many species will become extinct before they are even discovered.

Putting aside the effects of climate change on extinctions, humans are directly responsible for many impacts. A short list includes predation by humans — such as how we go for the biggest animals and how we overfish and overhunt species. To date there has only been one detailed observation of a nonhuman animal overhunting. Chimpanzees in Uganda's Kibale National Park work together to catch prey, typically red colobus monkeys. Between 1975 and 2007, there was an 89 percent decline in the red colobus population, while the chimpanzee population rose by about 83 percent. The list of direct human impacts also includes pollution — which destroys ecosystems and causes diseases, but which also leads to species and pests becoming resistant to poisons and pollution and creating further ecological problems. It also includes introducing invasive species — which can completely overrun native populations and permanently alter environments.

At a talk I heard by Cornell University's Christopher Clark in November 2012, he discussed "the ocean global commons" and the devastating effects of our acoustic footprint underwater, largely due to commercial exploitation. My mind was blown by how much large-scale damage we do to marine life just by the noise we make. As with many of our impacts on nonhuman animals, most people are totally unaware of this because the damage is hidden from our direct view.

Given our transformative, and often destructive, presence on Earth, some scientists now propose that "humans have become the biggest force in evolution." They characterize this as "unnatural selection," but I wonder how useful it is to try to distinguish "natural" and "unnatural" selection. After all, humans are an integral part of nature; we have a "natural history"; we are great apes. Using terms like "unnatural" only encourages an "us" versus "them" mentality, even though it's meant to wake us up to our unseen, profound impacts on nonhuman animals.

For instance, in a *New Scientist* article entitled "Unnatural Selection," Michael Le Page writes: "The Zoque people of Mexico hold a ceremony every year during which they grind up a poisonous plant and pour the mixture into a river running through a cave.... Any of the river's molly fish that float to the surface are seen as a gift from the gods. The gods seem to be on the side of the fish, though — the fish in the poisoned parts of the river are becoming resistant to the plant's active ingredient, rotenone. If fish can evolve in response to a small religious ceremony, just imagine the effects of all the other changes we are making to the planet."

Given all the ways that humans negatively impact other species, Le Page concludes, "It is no secret that many — perhaps even most — species living today are likely to be wiped out over the next century or two as a result of this multiple onslaught. What is now becoming clear is that few of the species that survive will live on unchanged."

Our effects on other species are wide-ranging and far-reaching, and we most likely understate the extent of our destructive ways. As with climate change, we often don't know or fully understand what we've done or the extent of our negative impacts. Even worse, we have no idea how to fix the ecological problems confronting us, whether we are at fault for them or not.

Unwilding: The Roots of Alienation

If we did not unwild, there would be no reason to rewild, and we need to reverse this distancing and destructive devolution. If by "rewilding our hearts" I'm naming our open and compassionate connection to nature, then unwilding refers to the opposite: It's the process by which we become alienated from nature and nonhuman animals; it's how we deny our impacts and refuse to take responsibility for them; and it's how we become discouraged and overwhelmed, and thus fail to act despite the problems we see.

Many, perhaps most, human animals are isolated and fragmented from nonhuman animals and other nature, and so we become alienated from them. The busy-ness of our days, the concrete and steel of our cities, the buildings in which we spend the majority of our work and school lives — all this unwilds us and erodes our natural connection with nature. German psychologist Erich Fromm called this innate connection, this love of life and living systems, "biophilia." Renowned biologist Edward O. Wilson later defined his "biophilia hypothesis" as "the connections that human beings subconsciously seek with the rest of life."

Yet our modern world undermines this constantly. It unwilds us. We experience alienation from nature when we learn about, or participate in, the wanton killing of wild species, when fields and forests are clear-cut and paved over for suburban developments, and when ecosystems are ruined by pollution or other human impacts. We experience firsthand our separation from nonhuman animals when we keep them in cages in zoos. And we instill alienation from nature in our children by teaching them primarily indoors at desks and in front of computer screens. Alienation flows from the belief that humans are superior to all other animals and that we are meant to dominate other species and use the Earth solely for our benefit.

Many people run within very narrow worlds, so that they never feel and can't imagine that all people are connected and that human life is inherently dependent on nature's health. It can seem ludicrous to claim that what happens in New York City or Boulder, Colorado, really does influence what happens in other parts of the world. After all, from our perspective where we live, other parts of the world don't seem to influence or impact us. Particularly in America, though this is true everywhere, some people are fortunate enough to live very comfortably, and they face few (or fewer) problems in how they live, and so the serious problems that exist elsewhere can seem distant. They aren't experienced immediately and directly, and they are easy to ignore. The welfare of faraway places we've never visited, and of creatures we've never seen, doesn't seem to concern us.

As we unwild, we lose compassion and empathy for other beings and for nature as a whole. We do not understand that landscapes are alive, vibrant, dynamic, magical, magnificent, and interconnected. Nevertheless, unwilding can often lead us to experience a deep sense of loss for the connections to nature we are missing, even if the sense of what we're missing is vague. Glenn Albrecht, professor of sustainability at Murdoch University in Perth, Australia, has coined the word "solastalgia" to describe "the distress caused by the lived experience of the transformation of one's home and sense of belonging and is experienced through the feeling of desolation about its change." We experience solastalgia when we erode our relationships with nature and other beings.

Homo denialus

If alienation from nature is perhaps an unintended consequence of modern life, people also engage in a more deliberate form of unwilding: denial. As *Homo denialus,* we readily "see no evil, hear no evil, or smell no evil." If we open our senses even a little bit to our surroundings, it is impossible to miss what's happening and the dire consequences of our actions, but we can get mired in "magnificent delusions" (to quote political scientist Husain Haqqani). We ignore and redecorate nature in incredibly self-serving ways, as if we are the only species that matters, and we turn a blind eye to the suffering this causes.

An old Chinese proverb warns us that closing our eyes does not ease another's pain. Claiming ignorance and denying what is happening do not make the destruction stop. When did we begin ignoring nature? Why did we start ignoring our need for untainted and healthy food, clean water, clean air, and reasonable shelter? How did we become so disconnected from nature and an understanding of basic ecological processes? What allows us to tolerate human-induced losses in biodiversity? There can be many reasons for denial. Some hide behind the claim of human exceptionalism, and so they ignore the suffering of other animals because they think those animals are less than us and don't matter. Some deny the destruction that humans cause in the natural world in order to avoid having to take responsibility for it.

Also, in the hustle and bustle of modern life, it is easy to simply become too distracted and to ignore nature as we run here and there, not even knowing why we're doing what we're doing. We are removed from the larger impacts of our own daily lives, and so we lack any meaningful appreciation of our destructive ways and their wide-ranging consequences. To a degree, this distraction is self-inflicted and self-serving; we ignore nature because it's convenient to do so. We choose to live in oblivion and deliberately not know and not feel what we're doing. We prefer detachment and ignorance rather than to feel and share the pain experienced by other animals and nature.

However, denial is hard to maintain when the evidence of trouble exists all around us. Some persist anyway. They deny solid science and what is happening right in front of our eyes. This has real consequences, which Michael Specter makes clear in his book *Denialism* and Richard Oppenlander in *Comfortably Unaware*. Renowned conservation biologist Michael Soulé calls the climate change deniers "morbidly ignorant." These are strong words from a world-famous scientist, one who is often called the founder of conservation biology. I agree with him, and it's one reason why I have written this book — to help undo these unwilding mechanisms so we can start to rewild our hearts before it's too late.

Combating Slacktivism: No More Excuses

Most readers of this book probably already recognize and acknowledge the very serious ecological problems facing us. What's harder is actually doing something. Particularly when we are faced with unrelenting bad news about global issues like climate change, which make our individual efforts seem inconsequential and useless, it is easy to throw up our hands and give in to despair and inaction.

But everyone must do whatever they can. We cannot be slacktivists. Many people talk about making the world a more compassionate place for all beings and "living green," but one recent US survey found that "fewer than 10 percent use any environmentally friendly products or curb household consumption." This is slacktivism: talking about an urgent problem that needs to be fixed without walking the talk and actually doing something about it.

Doomsday thinking is another form of unwilding. It fosters alienation and isolation, from nature, nonhuman animals, and one another. Feeling overwhelmed, we may retreat into our individual lives, never realizing how keenly this disconnection is felt by others. What we do *does* make a difference, and rewilding our hearts is about fostering and honoring our connections to one another and all life. We don't need to have all the answers to the world's problems in order to make the world a better place where we live right now. But we do need the determination to push aside the excuses and rationalizations for inaction whenever they come up.

There are dozens of excuses for negativity: thinking that there's nothing we can do to turn the tide; failing to consider future generations, who rely on our goodwill and efforts, even if we ourselves might not live to see the benefits of our actions; preferring our own immediate gratification or comfort rather than making sacrifices for another's benefit; outright laziness; and hierarchical speciesistic thinking that we are "higher, better, or more valuable" than other animals. Some might also cite religious beliefs or proclivities, or appeal to economics — that it's too expensive or time-consuming to care and make the necessary changes. Of course, politics gets in the way all the time: We know what's "right," but what's right isn't always politically expedient or in alignment with our faction's dogma.

Fear of and unfamiliarity with the outdoors can also get in the way. These are other forms of unwilding. We can be reluctant to embrace or connect with something that makes us uncomfortable, and some might fear that animals are dangerous or that they are dirty or carry diseases. People may avoid the rugged outdoors because they are afraid to get dirty, afraid of falling and being injured, afraid of insects and of encountering wild animals. However, whatever one's personal comfort level with nature, we can all work on nature's behalf, for nature is a common good we can all recognize. Rewilding our hearts doesn't mean becoming an "off-the-grid" survivalist, a radical "back-to-the-land" activist, or a hard-core outdoorsperson. It means, simply, acting with compassion and love for nonhuman animals and for the natural world that is our shared home.

We are also inconsistent in our caring. People are often outraged over specific incidents of animal cruelty — such as the massacre of forty-nine captive wild animals in Ohio in October 2011— but they remain unmoved by the slaughter of billions of animals for food and research, or the horrific and ongoing abuse of animals used for entertainment in zoos, aquariums, circuses, and rodeos. While not every situation is equivalent, rewilding means not selectively picking and choosing who to care for based on our own whims. As a comparison, it has been estimated that, each year, 37 to 120 billion farmed fish, 970 to 2,700 billion wild fish, and 63 billion farmed mammals and birds are killed for food.

Excuses for inaction are a dime a dozen: Life is too demanding, there's too much to do, money is tight, individuals really don't make a difference, someone else will take care of it, I'm not at fault . . . the list goes on and on. Clearly, bettering the lives of animals and healing our ecosystems are truly daunting tasks, and relatively few people in the world have the ability to make it their full-time job. Most people in the world are already doing all they can just to survive. Those of us blessed with good fortune often forget that the vast majority of people in the world, despite all their efforts, barely make it from day to day. And no one is completely free from the need to take care of their own welfare and that of their family. But that makes it all the more important that each of us do something and that we push ourselves to do as much as we can for as long as we can. Indeed, whatever our circumstances, rewilding is really a lifelong effort. We should think of it not as series of one-time actions that "do our part" for the environment but as a lens for viewing and remaking our life from here on.

Science and common sense tell us that we have made some

egregious errors that have created numerous lasting or long-term problems and that we need to change our ways. But science alone won't convince us to change. Science alone will not add compassion to the world or get people to do something positive for animals and their homes. Only rewilding our hearts will do that. Only by embracing rewilding will we avoid the alienation, denial, and complacency that undermines our efforts. Rewilding is a process that begins in each person's heart and expands outward, one that heals our own connection to nature even as it heals the wounds of our one and only planet. Rewilding can give legs to a new social movement and paradigm shift for much-needed change.

Rewilding Comes Naturally

Over the past few years I have had the pleasure of talking with the renowned psychologist Aubrey Fine, who is keenly interested in the nature of human-animal relationships. He once asked me a simple question, "What do we need to do to get people to stop and smell the roses?"

As I told Dr. Fine, to me it should be easy. We are born biophiliacs who are inherently drawn to the natural world. This attraction is in our genes. People don't need convincing to enjoy nature. When we take a walk outside, we notice immediately how much better we feel. I would love to see brain scans of people as they rewild and reconnect with other animals and nature. I would not be at all surprised to see their reward and pleasure centers firing wildly.

In the next chapter, I describe rewilding in more detail, but at the very least this is what rewilding is all about: encouraging, honoring, and growing our inherent connection to nature and nonhuman animals. After this chapter's difficult summary of the problems we face, I want to end on a positive note by emphasizing that the solution, connecting with nature and interacting with compassion, is based on who we are.

For instance, the distractions of daily life can make it very easy and convenient to unwild. We become, in some sense, out of balance and disconnected within ourselves. As we rewild, we restore balance and our sense of connection. We achieve a sort of homeostasis and feel good once again — it's like coming home to a comfortable place. From this place of connection, it's easy to see and do "what's right." I love something that Ernest Hemingway wrote years ago in *Death in the Afternoon*: "About morals I know only that what is moral is what you feel good after and what is immoral is what you feel bad after." I don't see Hemingway's view as simplistic at all. Rewilding is the same: You know it by how much better it makes you feel.

The challenge, of course, is to maintain rewilding and these feelings as a daily routine. But all sorts of evidence continues to emerge that kindness, compassion, caring, and cooperation are part of our evolutionary inheritance. For instance, research has shown that the roots of fairness and cooperation can be found in infants as young as fifteen months of age and that egalitarian instincts emerge in early childhood. Mounting research in evolutionary biology shows that groups comprising individuals who work together do better than groups comprising individuals who do not. This supports an evolutionary theory called "group selection," in which cooperative groups provide an evolutionary advantage, in contrast to the more well-accepted theory of evolution that argues that natural selection operates on individuals. Group selection as a driving force in the evolution of social behavior has generated more support over recent years, including that of renowned biologist Edward O. Wilson, who for many years strongly opposed it.

I've often said that, across cultures, humans are really much nicer than we ever give them credit for. It's a relative few who wage wars, kill people, and harm children, and they get in the news. This is true for nonhuman animals as well. Kindness is seen in a wide variety of species. I tell more stories of animal compassion later, but I love this story about Gus, a dog, which was shared with me in an email:

> Gus, as most dogs are, was fiercely territorial about his yard. Any animal that showed up would be promptly and definitively chased off...except for one day: It was a scorcher of a day in Northern Alberta, and Dad and the family were sitting at the kitchen table with the door open, hoping for a bit of a breeze to cool things off. Suddenly, at the door appears a stray, starving dog. Right behind is Gus, who gives the stray a gentle nudge forward. He then walks in front of the stray animal and leads it to his food dish. Gus put his paw in his dish and looks at his owners as a sign to please fill this with food, which they did. They watched in awe as Gus stepped back and let the dog empty his food bowl. After Gus was confident the animal had a full belly, he then chased the dog out of the yard.

For Gus, and for all of us, there is always room for compassion. And there's mounting evidence that being compassionate is good for our health and longevity. Further, studies show that people are willing to pay to enjoy all sorts of nature. Conservation psychologist Susan Clayton at the College of Wooster reports that people value and are willing to pay "for scenic beauty; for diversity of animal species; to protect habitat for the giant panda; to reduce invasive plant species; to protect biodiversity in urban areas," and more. She continues, "Perhaps the most striking finding is that people are willing to pay for aspects of the landscape that they would never be able to benefit from personally. This idea has been described as 'existence value': We value the mere existence of things like the Grand Canyon, the Arctic National Wildlife Refuge, and many other instances of nature that remain relatively unaffected by human activity."

Rewilding only asks us to do what comes naturally and what feels right. And despite our current problems, we have an incredible amount of global momentum. More and more people realize that they too can make a positive difference in the lives of other animals, other humans, and in the health and integrity of our landscapes. The rewards for this work are felt immediately, and the effects ripple in all directions, influencing others to do the same. Each time we nurture the seeds of compassion, empathy, and love, we deepen our respect for and kinship with the universe. All people, other animals, human communities, and environments benefit greatly when we develop and maintain a heartfelt compassion that is as reflexive as breathing. Compassion begets compassion.

2. Compassion and Coexistence Mean It's Not All about Us

If we don't always start from nature, we certainly come to her in our hour of need.

— Henry Miller

In the traditions of the many different Native peoples of North America, animals are almost universally seen as equals to humans on the circle of life. The word "circle" is especially appropriate, for all living things, animals and humans alike, are part of a great circle. No part of that circle is more important than another, but all parts of that circle are affected when one part is broken. In the eyes of the Native American, animals are all our relations.

— Joseph Bruchac

To rewild, we need to make compassion, empathy, and peaceful coexistence social values. In *Humanity on a Tightrope*, Paul Ehrlich and Robert Ornstein make a strong case that one way to solve the numerous challenging predicaments in which we find ourselves is by "quickly spreading the domain of empathy." We live in a messy, complicated, frustrating, demanding world, and it is impossible to do the right thing all of the time, however we define it. Compassion is the glue that holds ecosystems, webs of nature, and circles of life together. Compassion also holds us together. We are an integral part of many beautiful, awe-inspiring, and far-reaching webs of nature, and we all suffer when these complex interrelationships are compromised. This is not a new idea. Interconnectedness and interdependence are central to a Buddhist perspective as well as to all indigenous spirituality. Often, we imagine ourselves standing apart from and above nature, as what Neil Evernden calls "natural aliens." Yet Pat Shipman asks poignantly, "If our species was born of a world rich with animals, can we flourish in one where biodiversity is decimated?"

Rewilding names the revolutionary paradigm shift we need in how we interact with other animals and Earth. Since it has to begin somewhere, we might easily start with the question, "Who do we think we are?" Indeed, much of this book deals directly or indirectly with this truly daunting question. In this chapter, I address the "It's all about us" paradigm that guides so many of our decisions. Too often, we cause ecological problems and animal suffering because we think of ourselves as the only beings who matter.

Taking Exception with Human Exceptionalism

Rewilding aims to undo two particularly damaging and self-destructive attitudes: one, that humans as a species are "better" and "more important" than other animals, and two, that humans are somehow immune from the incredible destruction that our activities are causing to the planet. Chapter 1 hopefully dispels any lingering doubt about the latter. So what about the former?

No matter what we otherwise claim, most of the time we justify any negative impacts by engaging in rampant anthropocentrism. On some level, whatever we do is okay because "it's all about us." This is speciesism, plain and simple, and it's guided human thinking for a long time (as Gary Steiner relates in *Anthropocentrism and Its Discontents*). Yet it's a selfish attitude that not only is ethically bankrupt and harmful to our own welfare but is bad biology and science.

First of all, as anthropologist Barbara King points out, it is not "our human evolutionary birthright to be the dominant animal in the landscape.... The Anthropocene is not a natural outcome of our evolutionary trajectory." We remain, as we always have been, one of a gang of many species who give and take from one another and from the habitats in which we live. We remain part of the natural world, part of a collective in which all animals play important roles. Ecologically speaking, humans are far from the most important species. In this collective, "entangled empathy" (to quote the 2014 book by Lori Gruen) is a necessary ingredient. We must understand what other individuals, human and nonhuman alike, want, need, and feel so that the collective is cared for, not just one species at the expense of countless others.

Without question, the human species evolved traits and abilities that allowed us to become the dominant animal, but as any grade-school child will tell you, the ability to dominate others is not what makes someone the "best" or most deserving. Bullies dominate. Of course, by definition, the human species is different from other species, and in some truly unique ways, but that doesn't mean we are the template against which all other beings should be measured, nor does it mean that our selfish interests should trump those of the other species with whom we share our planet. Rewilding stresses the importance of reciprocal interconnections among species, who depend on one another to maintain balance in our complex world. Even if this were not so, it's equally true that other animals are unique, special, and "exceptional" in their own ways. When other animals do things that humans can't, does that make other animals inherently "better," more important, or more valuable than us? Human exceptionalism fails because we're one of the gang, part of the complex, magnificent, and daunting puzzle of an interconnected Earth.

Still, people sometimes ask, aren't the attributes that make humans unique "better" than the attributes that make other animals unique? Aren't our thoughts more complex, our emotions richer, our morals higher? Aren't we, on some level, still qualitatively "better" than other animals, even if, as animals, we are still connected and, to wildly varying degrees, related to other species? It's very seductive to think so, but no.

As I've written in other places, research over the last fifty or so years, but especially recently, has undermined one point of human exceptionalism after another. Nonetheless, some people ignore or deny what we really know about the lives of other animals. We know, for instance, that other animals are conscious, rational thinkers who communicate using a wide variety of complex signals. We know that many animals have rich and deep emotional lives that range from unfettered joy and happiness to devastating sadness and heart-wrenching grief. Some species can also lie, make jokes, laugh, cry, console, cooperate, and organize a complex jailbreak from their zoo cages that includes picking locks and fooling zookeepers for weeks. We know some species use and make tools (including fish and amphibians) and some engage in symbolic communication; dolphins and apes have even learned to "talk" with us using systems of human-created symbols and signs. We also know certain animals make moral, compassionate choices to help others or avoid causing pain to others, even if they have to make sacrifices for themselves. Rats, for example, will choose to help another rat in need rather than eat chocolate themselves. Of the more complex emotions and morality, very few traits cleanly separate us from other animals. We simply cannot know if other animals hold spiritual beliefs, but other animals have been observed acting in ways that suggest they possess a sense of awe and wonder about the world. With the aides of language and writing, along with our big brains, humans may indeed be capable of the "most complex" thoughts, but an ever-growing database of research cautions that we honestly cannot say that our thoughts are *always* more complex or that our emotions are more deeply felt than those of other animals.

I find it amusing, in fact, to try to create lists of the things that actually make our species unique — those attributes and activities that, as far as we know, no other species displays. So, what does set us apart? Here is a short sample of unique species attributes:

We are the only species who cooks food.
We are the only species who makes and uses fire.
We are the only species with a written language.
We are the only species who engages in mass murders and wages global wars.
We are the only species who has farting contests or tries to light farts (although I have shared my home with

dogs who sometimes seemed to be trying to outdo one another).

We are the only species to hold bake sales.

We are the only species who wears raincoats, hats, or running shoes.

We are the only species who uses condoms.

We are the only species who tattoos themselves.

We are the only species to use modes of transportation other than our bodies.

We're the only species who uses computers, Twitter, or Facebook.

We are the only species who has tried to leave the planet.

And we're currently the only species whose lifestyle and reproductive capacity impacts and damages every ecosystem on our planet.

Making Moral Choices: First, Do No Harm

Most of my career has been spent researching and publicizing the amazing abilities of nonhuman animals. However, this has not been done as a way to somehow denigrate humans while elevating other animals above us. To me, comparisons, hierarchies, and qualitative judgments like "better or worse," "smarter or dumber," or "higher or lower" are highly misleading and "bad biology." Ultimately, rewilding our hearts and reconnecting with nature and other animals means striving for peaceful coexistence no matter how sentient or smart we believe nonhuman animals to be. Rewilding is about the moral choices we make. It's about valuing all life as inherently good and worthy, without qualification. It's about expanding our relatively closed human clubhouse to incorporate all of Earth. Our fellow creatures depend on our goodwill for their survival, and we cannot continue as we have, willy-nilly, without taking into account their presence.

Rewilding is a pact we make with nature: to do no intentional harm, to treat all individuals with compassion, and to step lightly into the lives of other beings and landscapes, including bodies of water and the atmosphere. What this means in real terms will vary depending on each situation and circumstance. It will be difficult, challenging, and frustrating to achieve win-win solutions all of the time. Few of our problems today lend themselves to perfect solutions. There are always trade-offs. All stakeholders may have to give up something, and some more than others. But if we set any lower standard than compassionate coexistence, we can be sure that everyone will lose eventually.

The truth is, coexistence can be difficult. Other animals can cause problems for us. In frustration, some people call certain animals "pests." They refer to them as a nuisance, vermin, or trash. Obviously, calling animals "pests" or "trash" conveys the unsubtle message that their welfare isn't as important as ours. These labels make it easier for us to justify harming and killing these animals. But the animals who trouble us are not the real problem; they are only the messengers for a situation that is unworkable, out of balance, or destructive. As my colleague Philip Tedeschi, who heads the University of Denver's Institute for Human-Animal Connection, notes, "We have created the human framework for the problem, and then blame the animal."

Thus, the flip side to the imperative to do no harm is to take responsibility for our role in every situation. We often have unrealistic expectations, or we define our needs and build our communities such that animals will inevitably become a problem. This reminds me of how some zoo administrators call animals who are not part of their captive breeding program "surplus" animals, and then they kill these animals because they are of no use to the zoo. For example, in early 2014, the Copenhagen Zoo killed a young healthy male giraffe named Marius because he couldn't be used as a breeding machine, and later four lions, including two cubs, were killed at the same zoo so that a new male could be introduced to the remaining females. It's a perversion of logic and morality to breed animals to "save" species only to kill those same animals when they become too inconvenient to care for.

In his book *Respect for Nature: A Theory of Environmental Ethics*, which should be required reading for everyone working in the conservation sciences, philosopher Paul Taylor outlines four broad rules of how we should interact with animals and other nature. These duties include the rule of nonmaleficence, the rule of noninterference, the rule of fidelity, and the rule of restitutive justice. This means: We must strive to do no intentional harm, to treat other animals with respect, to be faithful to other animals, to not interfere in their lives, and to give back to nature and other animals what we take away. Taken together, these duties represent a powerful and sweeping argument for taking seriously the lives of individual animals. We must be committed to looking for the most humane solutions to the problems at hand and to require of ourselves compelling and forceful arguments when overriding these strictures. The interests of all stakeholders must be taken into account, and a hands-off approach should always be considered the best thing to do.

However, besides living up to our own moral standards, we should treat nonhuman animals with compassion because, as I note above, we know they possess empathy and compassion as well. Research now proposes that morality and compassion are "rooted in our biology rather than our intellect." That is, these responses are, in part, evolved and innate, rather than only learned. We also know that these feelings readily cross species lines. Many different animals display compassion for other species, so it should be natural for us to call upon compassion to alleviate the suffering of others. Richard Foster, editor of the *Daily Kumquat*, wrote me a most moving comment: "The blind eye we turn to the suffering of animals is probably the greatest example of cognitive dissonance in the world."

People often send me stories and examples of animal "odd couples," those individuals of different species who display compassion for each other and form unlikely friendships. These are species who "shouldn't" develop strong and enduring bonds, but they do. These friendships demonstrate that joy, love, empathy, compassion, kindness, and grief can readily be shared across species lines, even between typical predators and prey. Examples I've heard of include a cat and a bird, a snake and a hamster, a lioness and a baby oryx, a cheetah and a retriever, a lion and a coyote, a dog and a deer, a goat and a blind horse, and even a tortoise and a goose. Of course, the best examples of shared caring between different species are those close and enduring relationships we humans form with our companion animals and with those nonhuman animals with whom we work closely to rehabilitate and heal when they are in need.

Along these lines, Marius Donker, a retired Dutch psychologist, sent me this wonderful story:

> Your remarks on elephants reminded me of a short article in a Kenyan paper, when I visited Kenya around 1978. From what I remember the story went like this: Enjoying the sunset a British lady ventured outside the perimeters of a lodge near Nairobi into the high grass. A few hundred meters from the lodge she sat down, relaxed and fell asleep. She woke up surrounded by an elephant family that gently touched her body. Scared and not knowing what to do, she kept her eyes shut and feigned being dead. After more sniffing and touching, the elephants covered her carefully with collected thorny branches and left. She was found by a search party a few hours later. I always wondered if these elephants knew whether she was alive or not. Anyway, it was a compassionate action to protect this tourist from predators.

Rewilding: A Silent, Spiritual Revolution

Ultimately, rewilding is not merely a logical argument or a moral imperative. It is an expression of love. It is our response to the unspeakable wonder and amazement of creation itself. The new social revolution that centers on rewilding our hearts is a silent and spiritual movement that honors simplicity. In *The Great Work,* the late theologian Thomas Berry stressed that our relationship with nature should be one of *awe*, not one of *use*. Individuals have inherent or intrinsic value because they exist, and this alone mandates that we coexist with them. They have no less right than we do to live their lives without our intrusions, they deserve dignity and respect, and we need to accept them for who they are.

Almut Beringer, from the University of Prince Edward Island in Canada, describes this as a "spiritual handshake." We all need to take care of ourselves, and we also must take care of others, and so we should approach every situation as a balance between personal self-interest and service for the common good.

Others call this perspective deep ethology, which balances our studies of animal minds with respect for their hearts. Deep ethology rewilds our hearts and helps us overcome speciesism. Leslie Sponsel, an anthropologist at the University of Hawaii, describes this same idea as "spiritual ecology" (in his book of the same name). Through simple activities like taking a morning walk in nature, he proposes we can restore our "ecosanity." In a similar way, my esteemed colleague Michael Tobias, an ecologist and award-winning filmmaker, rocked the souls of everyone at a 2012 symposium on biodiversity and conservation when he asked, "How hard is it to give a crumb to a bird or a child?"

Given the severity of the problems facing us, it may seem crazy to suggest that simple, tiny acts of caring could lead us out of our troubles. But this is what I have come to after thinking about the nature of human-animal interactions for decades. Every act matters, and every act reflects the deeper attitudes behind it. Those attitudes influence all our actions. Nothing will ever get solved if we don't reconnect with the deep richness of diverse and magnificent nature in intimate, personal ways. Like feeding the birds, our everyday interactions with nature and animals should engender moving feelings of warmth and care. This is why I say that rewilding is not a program or a list of actions; it is an intuitive feeling of connection, one that continues to grow and become ever-more inclusive. Rewilding is an emotional, moral, and spiritual guide for evaluating all our interactions, big and small, with other animals and with Mother Nature. It is a felt truth of honesty and engagement. When I think about connecting and reconnecting with nature, I get a warm feeling that fills me with hope and inspiration that I want to share with others.

Henry Miller's quotation with which I begin this chapter rings true for me and for many others. So does University of North Carolina psychologist Marc Berman's views that nature can improve brain performance. But why is this so? Why do we go to nature for guidance? Why do we feel so good when we see, hear, smell, and if possible, touch other animals, when we look at and touch trees and smell the fragrance of flowers, when we watch rushing water in a stream or the waves of the ocean? We often cannot put in words why nature has such positive effects on us — why, when we are immersed in nature, we become breathless. We can place a hand on our heart and feel our heart rate slow down in the presence of nature's beauty, awe, mystery, simplicity, multiplicity, and generosity. This is why rewilding is a "silent" revolution. Rewilding occurs in the breathless awe of our encounters with nature. But just because we cannot translate our experience into words does not mean that nature does not have deep and long-lasting positive effects. Clearly she does.

Perhaps our inability to express nature's effects simply means that the feelings evoked are too deep, touching our souls. There are no words deep or rich enough to convey these feelings. We usually feel joy when we know that nature is doing well and feel deep sorrow and pain when we perceive that nature is destroyed, exploited, and devastated. I ache when I feel nature's wisdom being compromised and forced out of balance.

My primitive brain remains closely tied to nature. Rewilding is just that: the sheer joy we feel when nature is healthy, the joy we feel when we are embedded in nature's mysterious ways. It is the peace that occurs when the distance collapses and we feel at one with nature and all her creatures.

Science Alone Is Not Enough

When we make decisions that damage the environment or harm animals, it is rarely because of a lack of knowledge and concrete data. Rather, losses to biodiversity, inadequate animal protections, and other negative impacts are typically due to problems of human psychology and social and cultural factors. Science alone doesn't hold the answers to the current crisis nor does it get people to feel compassion or to act differently. As historian Lynn White wrote in his classic 1967 essay "The Historical Roots of Our Ecological Crisis": "More science and more technology are not going to get us out of the present ecological crisis until we find a new religion, or rethink our old one." More than four decades later this claim still holds: We do not need more science. We need a new mind-set and social movement that is transformational and centers on empathy, compassion, and being proactive. By rewilding our hearts, we focus on building strong and intimate connections with nature, and these experiences are essential for effective social change. This is deep work. Rewilding reconnects us with wildness, not just "the outdoors." I always like to keep in mind what Kathleen Dean Moore, a philosophy professor at Oregon State University, reminds us: "There is a difference...between the call of the *outdoors* and the call of the *wild*."

Along these lines, Michael Shermer writes, "The majority of our most deeply held beliefs are immune to attack by direct educational tools, especially for those who are not ready to hear contradictory evidence. Belief change comes from a combination of personal psychological readiness and a deeper social and cultural shift in the underlying zeitgeist, which is affected in part by education but is more the product of larger and harder-to-define political, economic, religious, and social changes." Echoing this, Joe Zammit-Lucia says poignantly, "Conservation is all about people." However, let me also stress that I agree totally with Dave Foreman, who notes in his important book *Take Back Conservation* that we must always keep nonhuman animals first and foremost in our deliberations and behavior. While rewilding is primarily a human social movement, our efforts to protect and save other animals should not be driven by our own interests. Indeed, focusing more on our needs would be exactly the opposite and wrong result. Rewilding conservation is to become less self-centered, not more so.

Of course, science and the results of empirical inquiries are important. This information can clarify the problems at hand and tell us what to do and how to act; science and research provide the critical practical information to make wise, effective decisions. But science doesn't move us to act in the first place. Despite all we know, environmental education has largely failed (for instance, see *The Failure of Environmental Education*, by Charles Saylan and Daniel Blumstein). Workable solutions to current and future problems depend on people from different disciplines first recognizing their common interests and then talking *with* one another, not *at* one another, and respecting each other's contributions. All the science in the world will not help us if we do not use its results. So, along with learning more, we must reach out to and listen to each other. We must respect the diversity of opinions, while convincing everyone that coexistence and compromise are the only attitudes that will lead us out of this mess and change things for the better. To convince people to right the wrongs, we have to better understand each other and create a common consensus that we can and *must* work to fix the things we break.

So, while scientific research continues and fills in the gaps in our knowledge, the precautionary principle mandates that we know enough right now to begin to right the wrongs. Our attitude should be "better safe than sorry." Whatever data we still lack, there are clear problems that need fixing. If our actions turn out to be ineffectual, we can learn from that and change what we're doing, but inaction is inexcusable.

Further, scientists themselves need rewilding to improve their work; rewilding makes for better and more useful science. By rewilding, we listen more closely to what animals need; we listen to the land. As we reconnect with nature, intuitively "reading" our world becomes as natural as breathing. Historically, science dismisses and cuts off such subjective impressions, this intuitive knowledge, but this attitude has often made scientists "tone deaf" to the animals they study. Too focused on proving what animals know, scientists don't hear what animals are saying: that they deserve caring, coexistence, and respect.

The perspective offered by David Haskell, an ecologist and evolutionary biologist at the University of the South, captures much of what rewilding is all about. Professor Haskell takes regular walks in a forest owned by the university, a place he calls "mandala." In this area he sits, watches, and listens. "Usually, if you stay here for a while, something is going to happen," he once said in a *New York Times* story, whose writer wrote: "As if on cue, a bird's cry pierces the cicadas' hum. 'There's a blue jay. There are the cicadas. There are the harvestmen crawling around.'"

This is rewilding in progress. Haskell noted, "Science deepens our intimacy with the world. But there is a danger in an exclusively scientific way of thinking. The forest is turned into a diagram; animals become mere mechanisms; nature's workings become clever graphs." However, a different understanding arises when you allow yourself to appreciate squirrels playing in the sun. "They are alive; they are our cousins. And they appear to enjoy the sun, a phenomenon that occurs nowhere in the curriculum of modern biology." The *New York Times* paraphrased Haskell's message: "Science is one story...true but not complete, and the world cannot be encompassed in one story."

I could not agree more. Science alone is not going to make people more compassionate, and conservation is all about getting people to work hard to save animals for the animals' own sake. This is the aim of the growing compassionate conservation movement, which involves people with wide-ranging interests — both academics and nonacademics — working hard together to balance the well-being of people, individual animals, and the health and integrity of landscapes and ecosystems. Academics, advocacy, and activism go hand in hand.

Nature and Compassion Are Good for Us

One afternoon as I was working on this book, I heard a thump on my office window and saw a small bird, a male chickadee, fall to the ground. He was stunned, and immediately two Steller's jays began attacking him. I ran outside, chased the jays away, picked up the tiny bird, carried him into my office, and placed him on my desk under a warm light. He looked terribly frightened and was panting. He tried to fly but couldn't. Slowly he recovered and began flapping his wings more vigorously. I took him outside, held him in my palm, and after a few minutes, he flew away as if nothing had happened. This brief rehabilitation really moved and rewilded me. I felt connected to the bird, and this small act of compassion filled me with warmth.

In writing this plea to rewild our hearts, I will appeal in one more way to our self-interest. Our alienation from other animals and nature stills and kills ours senses and our hearts, and we don't even realize how numb we have become until we witness the beauty of nature and the wonder of life: events as simple as a squirrel performing acrobatics as she runs across a telephone wire; a bird alighting on a tree limb and singing a beautiful song; a bee circling a flower; a lizard doing push-up displays as if to say, "This is my home"; or a child marveling at a line of ants crossing a hiking trail. In these small moments we feel our inherent connection with nature. Peaceful coexistence with other animals and other nature is essential, and our own physical and mental health are tightly bound with theirs. Recent books that highlight this include *Spiritual Ecology* by Leslie Sponsel, *Psychology for a Better World* by Niki Harré, and *Birthright: People and Nature in the Modern World* by Stephen Kellert.

While research can struggle to quantify the effects, and specific benefits are sometimes debated, studies are increasingly showing that connecting with nature is, in itself, good for our health. For instance, we've learned that young people who backpacked for three or more days showed higher creative and cognitive abilities than people who didn't. People in hospitals who can see natural landscapes recover faster than those who don't. Connecting with nature has also been shown to be helpful in reducing stress. Some benefits of being out in nature for youngsters include enhanced creativity and problem solving; enhanced cognitive abilities; improved academic performance, nutrition, and eyesight; and improved social relations and self-discipline. There is little doubt that future research will show even more ways that spending time in nature improves our physical and emotional health, and these good feelings will spill over into increased compassion and empathy for others, human and nonhuman alike.

Similarly, research is showing that, as I often say, compassion really does beget compassion. Surely, no one can be against a more peaceful and compassionate place to live. Rewilding our hearts — letting wildness coarse through our hearts and heads and impel us to act as the good and caring beings we truly are — is one way to create a more peaceful, compassionate, and just planet for all. As David DeSteno, director of the Social Emotions Group at Northeastern University, notes about recent research into the moral force of compassion: "[The] results are striking in that they demonstrate that compassion experienced for one person can instantaneously extinguish punishment for another. In short, His Holiness the Dalai Lama may be right: compassion may function to balance social systems so as to prevent escalating tit-for-tat aggression and downward spirals of prosocial behavior. Furthermore, this radiating ability may explain why sometimes any of us can be more forgiving toward someone than we ever would have predicted."

Given this, calling upon compassion to alleviate the suffering in nature can be seen as good medicine to heal our own human ills. *It is natural and healthy to be good, compassionate, empathic, and moral to other animals.* Do we need "more science" to be better or more compassionate? No, it is who we are. It is intrinsic to our common animal nature.

3. Making It Real

Hard Choices and Bottom Lines

No one wants to see polar bears, or pandas, become extinct. The question is, what will we do about it? It's inspiring to realize that, even as human behavior pushes more species to the brink, many people and organizations are passionately devoted to protecting them. Even Richard Nixon thought it was important enough to pass the Endangered Species Act. We don't really need to teach people to care. We just need to convince them to connect the dots.

— Susan Clayton

I'm constantly rewilding. It's a never-ending process. Until very recently, and for thirty-five years, I lived in the mountains outside of Boulder, Colorado. I always felt blessed to coexist with many magnificent and inquisitive nonhuman animals, some dangerous and some not. Cougars and black bears were among my visitors. Because I had to walk down winding stairs, and about fifty feet, to get from my office to my front door, I pretty much always felt a twinge of fear, especially at night or when I knew that a potentially dangerous animal had been recently seen, heard, or smelled. The fear invigorated me and made me appreciate that my home was situated in someone else's home, and I had better respect their presence. At any time they chose, I could wear out my welcome. It's the same sort of fear I felt in the Amboseli at Cynthia Moss's elephant camp when a lion rubbed against the outside of the tent when I was sleeping and pushed his body into my cot. But it's rather different from the fear and tension I have felt walking down a dark street in a bad part of a town. It is also different from the fear I have felt with a dog whose intentions I could not read. In my experience with a wide variety of wild animals, I've found them to be mostly indifferent to my presence, or they simply avoid me. But because these animals are wild, you never know what may happen if they decide I'm an intrusive or annoying "pest."

On many nights, I heard local animals trying to get onto my gated deck, gated because of them. One morning I noticed that a mother bear and her two cubs had succeeded in breaking the latch that held the gate closed. But most often, my brief commute between home and work involved less frightful encounters. I'd often see red foxes, chipmunks, squirrels, various birds, lizards, snakes, insects, and spiders, and I was always rejuvenated by the beautiful vegetation that surrounded my home. All of these feelings — of beauty and fear, of joy, wonder, and unpredictability — kept me in touch and attuned with the world where I chose to live, with who I was, and with who my nonhuman neighbors were. This was part of my ongoing personal rewilding.

Though my fears were usually niggling, and they never overcame my wonder, I knew that my anxieties were not unfounded.

As I was writing an early draft of this chapter, a handsome male bobcat came to my office door, stared in, and walked off and peed, totally unfettered. Soon after I almost tripped over a red fox who was sniffing where the bobcat had peed. On three occasions I wound up so close to a cougar I could have shaken paws with him. One hot summer day, I almost tripped over a large male as I was walking backward and telling a neighbor to watch out for the feline predator. When I turned to walk home, I could smell the cougar's breath and hear his panting, and his piercing eyes seemed to ask, "What in the world are you doing?"

Once, as I was getting ready to leave for the airport, a huge male black bear camped out on my deck and sat on the cover of my hot tub. He wouldn't leave, despite my nice requests and reasonable explanation that he was making me late for my flight. Later, when I returned home from my trip, there was a large pile of bear poop right at my front door. In June 2012, as I was riding my bicycle up the steep dirt road to my house, I was chased by a young bear cub, who surely could have caught me, but chose instead to stop chasing me about ten feet from my car. I was genuinely scared, but after my heart rate and breathing returned to normal, I honestly thought he was playing with me. Perhaps he was letting me know who was boss of our shared mountainside. Two months later, I met him once again when he began hanging out near my office. We agreed to coexist but nothing more. A red fox also once nipped at my heels as I rode, and during a 2010 cycling race, a black bear casually strolled right in front of me, sauntered across the road, and sat down. Naturally, I stopped, ceded him the right of way and chose to enjoy this unexpected reprieve from my arduous cycling up a 12-percent grade.

On a number of occasions, the first occurring in October

2011, the female black bear who had broken onto my deck also walked across my roof right above my bedroom in the middle of the night. She was no stranger to the neighborhood. I'd seen her often with her two cubs, and I'm sure she knew me and my habits well. The first time she sauntered back and forth across the skylight over my bed, I realized that this was not a good situation for either of us. The bear would have been just as unhappy as I if she fell through! I watched her silhouette, listened to her thumping about here and there, and planned an escape route, but there really was none. So I kept quiet, prayed the skylight would hold, and soon she was gone. The next times she came back, I was much more relaxed about her presence above me.

Do I feel lucky to have met these wonderful beings? Yes, indeed! Does that make me want to meet these amazing nonhuman beasts up front and personal again? Nope. Each time, I was lucky to get away unscathed, and I know it. But I also recognize that I made the choice to move into their living rooms in the first place, and I did so without their invitation. This feeling humbles me.

There's nothing better than a safe and peaceful home. That's what I want and need, and it's what other animals want and need. I liked to think of my mountain home and the beautiful land surrounding it as one of many peaceable kingdoms fostering heartfelt, compassionate coexistence. Animals want and need to live in peace and to feel safe, just like we do.

Of course, my life in the mountains had the usual ups and downs, but I never took my good fortune for granted. Every morning, when I woke up, I felt blessed. Animals abounded around my home because it was right in the middle of theirs. Why should I resent their presence, or my own occasional feelings of fear and of being an intruder? For my nonhuman neighbors, that's who I was — some guy who had invaded and now shared *their* home. This is true, to greater and lesser degrees, for all of us, no matter where we live. Nonhumans animals occupied the land before humans moved in, and nonhuman animals will continue to occupy the land, as they are able and as suits their own needs and survival. This includes, as it always has, the spaces in and around our own homes, towns, highways, airports, parks, farms, and so on. Why should animals have to move or be moved — or be killed, as they often are — because we find their presence a nuisance? Who are we to invade their homes and then tell them they have no right to live there anymore, and no right even to life?

As I was writing one morning, a red fox ran in front of my office window, stopped, peered in, and trotted off. He sort of paid attention to a chipmunk who was perched on a rock, but he seemed to realize that chasing this small rodent wasn't worth it — a mere appetizer — too small, too quick, and not enough calories to justify the effort. This male fox came by a lot, and I recognized him — he has a dark spot on his left hip. I have always wondered what he thought about me, seeing me in the same place at the same time pretty much every day. He wasn't afraid of me, nor I him, and occasionally I would walk outside, and he would come to within a few feet of me and sit down. I often spoke to him, and he would just sit there, patiently listening, saying little other than slowly sweeping his tail back and forth on the ground. One day I read to him what I'd been writing, and he seemed utterly bored.

Once he rolled over on his back as if asking me to rub his stomach. Of course, I would never do that. The fox was not my pet, and no good would come to him within his wild life by acclimating to my presence to that degree. I like to think we became friends, but I use the word loosely; the idea is perhaps self-serving. All I can say is that we certainly acted in a friendly manner, but there remained an unbridgeable distance between us; we were two very different animals with very different needs. I had moved into his ancestors' home years ago after someone built the house in which I lived, redecorating nature and his living room. He and the many other animals in the mountains always brought a smile to my face, but I never forgot that this was their home first. And I hope I was able to act in ways that kept their home a safe and peaceful place to live. This is the main goal rewilding strives for.

Redecorating Nature: Asking Tough Questions

As we rewild our hearts, one of the most important questions we face is what to do with the heartfelt healing concern for nature we are developing. As we build corridors of coexistence and compassion inside ourselves, how do we in real, practical ways "rewild the wild"? What can we do to coexist more peacefully and compassionately? What are the best ways to solve our many serious ecological and environmental problems? This chapter looks at several major areas that demand our attention and where rewilding is most needed. However, rather than debate specific solutions, I will mostly suggest how rewilding helps us rethink the problems and come from a more compassionate and empathic perspective.

Rewilding on any scale — whether we're discussing landscapes and biospheres or deeply personal experiences — means never ignoring nature. This is because wherever we go and wherever we live, humans are continually and inevitably redecorating nature, whether intentionally or unintentionally. All our buildings and roads, every human community, every vehicle, every energy source, affects nature and nonhuman animals. We can't help redecorating nature, and for some it's almost a human obsession. They can't leave nature alone. Nor is this unrelentingly intrusion going to change. Instead, it will only get worse. There are already too many of us, and more of us are arriving every day. As I wrote in chapter 1, our ongoing population explosion is one of the biggest environmental challenges facing us.

In short, our lives always impact nature, and it is difficult to live in our demanding world and not occasionally harm nonhuman animals, as much as we might try not to. This fact should keep us humble and nonjudgmental. We simply need to agree that we all must make every effort to minimize harm whenever and however we can. As this chapter makes clear, the best solutions are always a balancing act, a compromise between sometimes conflicting or competing needs. By agreeing to minimize harm to nonhuman animals and nature, we are agreeing to listen to the "voiceless" beings all around us and to proactively include their concerns in all our decisions. Increasing our awareness of nature, in itself, is the transformative first step.

Further, agreeing to minimize harm will lead us to focus on eliminating the obvious intentional harm that is currently being done and to confront it head-on. If we focused only on harm that is deliberate, easily avoided, or unnecessarily great, we could avoid much irreversible damage and make the world a better place for all beings. When we redecorate nature without compassion, we cause innumerable serious impacts on diverse ecosystems; we kill individuals, break up families, and decimate habitats. As nature becomes distressed, that in turn impacts us. Then, too often, our "solution" to such problems is to do more killing or to dominate nature further in our efforts to again be untroubled. To break this self-destructive cycle, we must agree that a nonkilling world is more important than our own uninterrupted comfort.

Knowing that other animals are sentient beings who experience unbounded joy and deep pain and suffering should help us care about what happens to them. This knowledge should help inspire us to want to minimize harm, though often it doesn't. Too often, we choose to remain ignorant of the many eyes watching us. Like my friend the red fox who regularly came by to say hello in the morning, animals see us, their lives matter to them, and they certainly notice when their lives are turned upside down and their homes destroyed. By rewilding our hearts, we pledge to treat all animal "visitors" as welcome sources of connection and inspiration. Minimizing harm and fostering coexistence are the default "best" choices. We need more compassion in the world. Yes, the human species leaves behind big footprints, but they can be footprints of compassion. There will always be trade-offs in deciding who and what to save and how to coexist. The tough questions facing us are incredibly difficult and frustrating. They ask each of us to think deeply and do the best we can for all beings.

The Eight Ps of Rewilding

Rewilding is an attitude. It's also a guide for action. As a social movement, it needs to be *proactive*, *positive*, *persistent*, *patient*, *peaceful*, *practical*, *powerful*, and *passionate* — which I call the eight Ps of rewilding. Because it comes from deep in our heart, the rewilding movement will be contagious and long-lasting and radiate far and wide. If we keep these eight principles in mind as we engage one another and wrestle with difficult problems, no one should feel threatened or left out. These are the principles that guide my own perspective on the issues in this chapter.

All of these attributes are interrelated and support each other. I've already discussed the need to be proactive. We can't wait for problems to get worse or for all the scientific data to be collected; we can't wait for a better, more convenient moment to act. We must act now with the information and understandings and abilities we have.

In addition to being proactive, we need to be positive and exit the stifling quagmire and vortex of negativity once and for all. Being positive and peaceful are extremely important, for many reasons. We need to talk up successes. Negativity is a time and energy bandit that thoroughly depletes us. We don't get anywhere dwelling in anguish, sorrow, pessimism, or despair. In a taxicab in Vancouver, British Columbia, I saw a sign in front of a church that really resonated with me. It read, "Make the most of the best and the least of the worst." Amen.

We also need to embody in ourselves the positive vision we hope to realize: a better future for all animals and our planet as a whole. If we present a negative picture, or act in negative ways, we can hardly expect others, especially those who remain unsure or unconvinced, to join us or do much at all. As we learn more and talk more as a society, I truly believe that we will be able to succeed in many of our attempts to right the numerous wrongs and bring more compassion, peace, and harmony to our world.

Plus, we know that being positive and hopeful are important

for getting people to care and to act. Concentrating on successes, on what works, is important for overcoming hopelessness. For example, it has been suggested that positive media — showing an elephant with her offspring — is more effective in getting people to right a wrong, such as elephant poaching, than negative images showing animals being violently tortured, such as elephants who have had their tusks ripped away. Scaring and shaming people does not work and can indeed backfire.

Almost by definition, rewilding means being open to the perspectives of all the parties impacted by a situation; it means bringing different points of view to the table, including all human points of view. As everyone's needs are shared, and impacts are seen, it should be impossible to be objective, indifferent, or complacent. Passion about our endangered ecosystems, about the well-being of nonhuman animals, and about our own welfare should help us get past indifference or a reluctance to act. As we articulate humanity's deadly and wide-ranging effects, passionate caring will help move us to make changes and avoid more harm. It astounds me that anyone could be indifferent when we are such an intimate part of the equation of what is happening right in front of our eyes and inside our hearts.

We also need to be persistent and stick like glue to our beliefs and ethical principles. Debate, differing opinions, and compromise are to be expected, but we can't give up or give in completely. By remaining positively passionate, by appealing to compassion with compassion — by "being nice" to those we disagree with, and talking *with*, not *to* or *at*, others — we can make the future better than the past. In this sense, if we think we are powerless, we will likely give up when things get tough. But we also need to persist by insisting on solutions that are themselves practical and powerful. Any solutions need to be sufficiently pragmatic and meaningful to make a difference.

Finally, patience may not seem like a virtue in these troubled times, but it surely is. There can be long time lags between taking action and seeing the effects. Getting consensus always seems to take too long, and compromises can sometimes feel like discouraging half-measures. Hopefully, we all agree that our troubled and wounded world needs a lot of compassionate healing right now, not when it's convenient. Ideally, we all share a compelling sense of urgency. But patience, taking time to smell the roses and to consider our actions carefully, is how we stay sane and maintain the energy to keep rewilding effectively each and every day.

Rewilding the Bottom Line, a.k.a. "Green Economics"

Wayne Pacelle, president and CEO of the Humane Society of the United States, argues we need a "humane economy," one that makes compassion and caring profitable. He correctly notes that cruelty is not only a moral problem but an economic problem as well. While Pacelle focuses on the lives of individual animals, others make a similar economic argument to view "conservation as investment." They have argued that in order to save the natural world from destruction, we need to put a cash value on it. Thus, one strategy for rewilding our business-oriented modern world is to rewild the bottom line, to rewild economics.

For instance, in an essay in *New Scientist* called "Costing the Earth," Fred Pearce writes:

> Welcome to the weird world of green economics, in which the value of ecosystems is being reduced — or elevated,

> depending on your perspective — to a matter of dollars, in a bid to save them from destruction. It makes a certain intuitive sense. Economics and ecology are intimately related, and not just in name (the "eco" in both has the same Greek root, *oikos*, or "house"). All economic activity is dependent on the environment: what would the timber trade be without forests, or fisheries without fish? And those are just some of the direct connections. Without a stable climate, water to drink and air to breathe, there would be no economy at all. But at present, the environmental factors that keep economies ticking over are almost entirely absent from economics itself. If they are acknowledged, it is as "externalities" that are not reflected in the prices of goods and services. A classic example is greenhouse gases: emitters emit them for free, and wider society picks up the tab.... Perhaps green economics is the least bad tool we have.

Obviously, you can't really put a price tag on nature. However, because so much of what we do or don't do comes down to money, it's a useful exercise that could form the basis of decisions that are made for different species and different problems. The price-tag approach might get us to reconsider our current choices by translating the "costs" and "benefits" of those choices into equivalent dollars and cents. By measuring the income of the businesses, like tourism, that rely on wilderness, we can quantify a relative economic value of wilderness itself. In the *New Scientist*, Pearce continues, "The biggest study yet, The Economics of Ecosystems and Biodiversity (TEEB), tried to value a range of important ecosystems. Coral reefs came out on top, worth up to $1.2 million per hectare per year, largely derived from tourism. The Amazon rainforest was reckoned to be worth between $6.5 and $13 billion a year as a carbon store alone. Even humble grasslands can be worth thousands of dollars per hectare, as protectors of water supplies and carbon stores. The TEEB study did not tell ecologists anything new, but it translated their knowledge into a metric that economists — and hopefully policymakers — can understand."

However, this economic line of reasoning is limited. As Thomas Berry noted, valuing nature for its human utility, and its monetary worth, only further objectifies the world. This is not rewilding. Nature and nonhuman animals have an intrinsic value that defies cost-benefit analysis, the same way we don't put a price on human life.

But also, appeals to economics often fall short of the sort of moral messages that get individuals to act. There are actions that are simply wrong no matter the monetary cost — causing intentional harm to animals, for example, and destroying their homes because it benefits us with little or no regard of the harm it causes them.

A study by researchers from the University of Groningen in the Netherlands showed that moral messages can get people "to go green," but they must appeal to the specific moral concerns of the individuals involved. The essay noted, "Previous research has suggested that morality can be divided into five main areas: harm and care, fairness and reciprocity, group loyalty, authority and respect, and purity and sanctity. Liberals place a higher priority on the first two areas than conservatives do, while conservatives are more likely than liberals to consider the last three areas important."

In other words, ethics, not money, typically inspires people to take action, and no single ethical reason moves everyone. This is why it's important to listen to, and ideally accommodate, everyone's needs and perspectives in any given situation. From a societal, business, or public policy perspective, putting a price on nature and nonhuman animals could be useful as a method of comparison, but economic arguments are probably much less persuasive on a personal level, especially when someone's profession or livelihood is at stake. For instance, government programs (such as in Sweden) that financially reimburse ranchers for livestock losses due to predators can sometimes persuade ranchers to tolerate those predators, rather than kill them, but not always. Money alone is not always an adequate incentive.

Compassionate Conservation and Restoring Ecosystems

Compassionate conservation has become a popular new phrase and mind-set over the past few years, one that has increasingly informed global efforts to conserve species and restore ecosystems. Compassionate conservation's central premise is that every *individual* animal's life counts, and so our attempts to repair ecological damage or save a certain species shouldn't, by design or accident, sacrifice or negatively impact even more life in the process.

In many ways, compassionate conservation represents exactly what rewilding our hearts is all about. It embraces compassion as an essential social value, one that should inspire us to right many of the wrongs for which we're responsible and one that should also guide all the decisions we make as we do. However, while it is impossible to argue against being compassionate, this approach presents us with some hard questions and may lead to some counterintuitive answers. What conservationists take to be gospel may need to be reexamined, and we may need to accept the limits of what humans can do, even if we think we have nature's best interests in mind and heart.

Whether we like it or not, these are some of the questions we need to ask, some of which undermine the very premise of conservation, such as: Should we even try to re-create or restore ecosystems when we really don't know how to accomplish it? Is a hands-off strategy more compassionate than trying to fix everything that's wrong? If we accept that we're an integral part of nature — "just one of the group," as ecopsychologist David Abram said in a recent talk, one doing very good and very bad things — should we just let ecosystems and their animal inhabitants evolve however it goes? Is it okay to kill some animals in the name of conservation or research — in other words, trading off individuals for the good of their own or other species or for ecosystem integrity? If so, to what degree? Should we focus our efforts on what seem to be the more soluble problems, even if these aren't the biggest problems? When do we pull the plug and admit defeat? Is it more compassionate to let some species go rather than try to do too much and spread ourselves and our resources too thin? How do we rebuild and adjust our human communities in order to create and maintain unobstructed natural corridors that allow species to revive? How can researchers, agencies, and policymakers cross disciplines and work together? When would animals and ecosystems be better without us?

For instance, with so many species currently endangered, we need to carefully consider which species we *choose* to save. Most often we focus on highly charismatic species, like pandas and polar bears, but in truth, these species often provide very little ecological value, and they are sometimes too far gone to have a chance of rebounding to viable populations without massive and ongoing human efforts. On the other hand, as we've seen, amphibians are disappearing at unprecedented and staggering rates, and their loss is bad not only for them but also for landscapes and humans. Conservationist Elizabeth Bennett points out that we really have not been very successful with protecting charismatic species, which leaves little hope for those who aren't eye-catching. Even young people, one report said, "place a higher priority on protecting colorful, showy critters than animals that blend into their surroundings."

Instead of focusing on species that appeal to us, shouldn't we focus on those dwindling species that are more or most important to the integrity of ecosystems overall? Consider honeybees. These amazing animals are responsible for pollinating seven of the world's ten most important crops, but a single bee can contain twenty-five different agrochemicals. Renowned conservation biologist Edward O. Wilson once said, "If all mankind were to disappear, the world would regenerate back to the rich state of equilibrium that existed ten thousand years ago. If insects were to vanish, the environment would collapse into chaos." Given this, should we put our limited resources into saving honeybees and amphibians and let wolves go if wolves can't survive on their own? As the amounts of time, energy, and money are strained as the number of problems grow, we're going to have to factor in more seriously the overall ecological importance of different species and make decisions that may favor animals about whom most people don't think or even know exist.

Obviously, we need to be selective about which species we choose to save and which ecosystems we choose to fix. This alone raises moral issues. However, I believe we might be more successful, and in some ways more compassionate, if we put our efforts into situations that are smaller and fixable, rather than trying to manipulate nature in fundamental ways. Some efforts are truly beyond our capabilities.

In truth, one way we ignore nature is with the belief that we can control nature. Rewilding our hearts also means accepting that, as with natural selection, life evolves as it must, and in this process, some make it and some do not. This is very sad, but also very true. We cannot undo the impact the human species has already had and restore nature as it once was; we can only move forward, proceeding from the way the world is right now. More and more researchers agree. In a 2011 survey, which questioned 583 conservation scientists, 60 percent said that criteria should be established for deciding which species to abandon in order to focus on saving others. Ninety-nine percent agreed that a serious loss of biodiversity is "likely, very likely, or virtually certain."

There is no avoiding further species loss. I wish this weren't so. Current strategies, sadly, were developed decades ago, before world populations and global consumption had skyrocketed.

As for how to go about recovering the species we do choose to save, a good approach is modeled by researchers working in Asia, who take what's called the Three Rs approach. The Three Rs are recognition, responsibility, and recovery, and the process entails recognizing the problem, taking responsibility for solving it, and putting species back on the path to recovery. Successful reintroductions, also called repatriations, include putting wolves back into the Yellowstone National Park ecosystem, golden-lion tamarins into Brazilian forests, black bears into the forests of Missouri, and red kites to areas of England. However, there are costs to these programs, as individual animals wind up dying "for the good of their species." A recent review of reintroduction studies involving captive animals showed that only about 30 percent of reintroduced animals survive. Humans are responsible for more than one-half the deaths due to shootings or car accidents. Of course, a lot hangs on the word "successful." If success is defined as increasing the number of animals who now live in habitats typical of their species, then there are some successes, but there also are significant failures.

Yet, methods are as important as results if the ethics of compassionate conservation are to guide us. My colleague Jessica Pierce stresses we need "ethics with teeth." That is, if recovering one species means killing a different species, we need to proceed very cautiously and perhaps even put such efforts on hold. For example, an effort to save endangered black-footed ferrets involves breeding and raising golden hamsters solely so they can be killed by the ferrets as practice prey. Over a few years, more than seven thousand hamsters were raised and killed only to serve as prey for captive-born ferrets. Is this a worthwhile trade-off? I don't think so, particularly since the effort was conducted without evidence that this practice hunting would translate in the wild or even that the ferrets would survive once they were reintroduced.

I don't pretend that these choices are obvious, simple, or even enjoyable. Yet I feel strongly that the imperative to do no harm should always come first, and it should make us cautious about actively trading one species for another, exchanging one set of lives we don't value for others we regard more fondly. History tells us it is a slippery slope when we sacrifice humane values for the common good. The real world demands that we come to terms with incredibly difficult and challenging situations, most of which we've created, and sometimes we must humbly accept that there is no solution that doesn't cause more harm in the process.

Seeking Nature: Ecotourism, Hunting, and Personal Enjoyment

One major reason humans preserve nature and wilderness, such as in state and national parks, is so that we can enjoy it. This positive impulse is central to personal rewilding, and visiting nature takes many forms. Some take walks or hikes in local woods for daily refreshment. Some take holidays to spectacular sights and landscapes, like Yellowstone National Park, the Grand Canyon, the Serengeti, Antarctica, and the Amazon. Some want to observe legendary animals in the wild, like elephants, penguins, bison, moose, bighorn sheep, and perhaps reintroduced wolves. Some spend lots of money to visit plush ecolodges, while others prefer rough outdoor adventures like rafting, mountain biking, and climbing. Some also hunt and fish; for them, enjoying nature also involves killing other animals.

Perhaps obviously, wildlife tourism impacts the wildlife and ecosystems being visited, though impacts vary depending on what people do. In this way, preserving nature for our benefit becomes another difficult topic that confronts several inherent contradictions when it comes to rewilding. First of all, if by rewilding our hearts we are in part attempting to lessen our impact on nature and nonhumans animals, then we should lessen our intrusions into nature. We should leave fewer actual footprints. And yet visiting nature, immersing ourselves in it, is crucial to what we most want, a deep and intimate connection to nature and nonhuman animals.

What to do? Continue to visit nature, but make a personal pledge to "leave no trace behind," as the slogan goes. We should dedicate ourselves to the least-impactful activities, even if that means lessening or forgoing activities we enjoy — like four-wheeling and hunting. This is one area where millions of individual choices can add up to a huge difference for wilderness and wildlife.

Then, looking at the bigger picture, we need to weigh the pros and cons of various types of wildlife tourism, and public land policy, and consider changes that improve our caretaking in the natural areas we have already set aside. This is a thorny topic that could be a whole book unto itself, and I can only touch on the main issues briefly.

On the one hand, ecotourism can have positive economic benefits that help preserve wilderness and save wildlife. Profits from tourism provide a financial incentive for local communities to conserve, rather than use or exploit, nature. In a very thoughtful 2012 essay, Ralf Buckley, who directs the International Centre for Tourism Research at Griffith University in Brisbane, Australia, notes that "many endangered species now rely to a startling degree" on wildlife tourism. Indeed, ecotourism has saved a number of populations of great apes, such as the mountain gorillas of the Virunga Volcanoes and Bwindi Impenetrable National Park in Uganda. Ecotourism has also been implicated in saving elephants in northern Kenya, as former poachers change their ways to protect these magnificent animals who attract people to their country — people who spend a good deal of money taking safaris to see the local wildlife. These sorts of programs are essential.

In a similar vein, some people argue that hunting animals has economic advantages, since hunters can spend a great deal of money on equipment, state fees, hiring outfitters, and so on. My home state of Colorado has a touchy-feely program called "hug a hunter" that encourages state residents to show appreciation for hunters and anglers because of their economic impact, which partly funds wildlife programs. However, in a very interesting and novel study, the Centre for Integral Economics analyzed revenues derived from British Columbia's grizzly bear resource. They found that grizzly bear viewing is conservatively worth $6.1 million annually in British Columbia, whereas hunting grizzlies brought in only half as much. The argument to maintain the hunt as an economic benefit when compared to ecotourism was not supported. Furthermore, legalizing the hunting of predators as a conservation strategy — on the theory that if hunters are allowed to kill a limited number of predators they will help conserve the species generally — actually has been shown to backfire, so it's wrong to conclude that sport hunters are "valuable partners in conservation."

As I mention above, in some situations, assigning cash values to animals and ecosystems will improve caring and compassion by giving us a financial incentive; we can quantify how peaceful coexistence is in our immediate self-interest. And yet, as I say, economics is a poor excuse for making moral choices. If we value wildlife and want to preserve it, then that is in conflict with killing wildlife for sport or enjoyment. Sometimes, rewilding our hearts will entail a "cost" to us. We may need to protect nature even when it's not financially advantageous or even when it means giving up activities we enjoy and once allowed ourselves to do freely.

Over the years, I've spoken with a lot of hunters and outfitters about the ethics and economics of hunting. Most of the time, hunters are very interested to exchange views and very respectful of my perspective, though I never know if any of them change their ways because of our talks. However, one argument usually gets hunters to stop and think. I remind them that their hobby, or their livelihood, causes pain and suffering by intentionally killing animals who have rich and deep emotional lives and who leave behind friends and families. Then, I ask the person if they would hunt and kill their own dog. "Of course not!" they usually respond with incredulity. But in the end, a dog and the wild animals people hunt are not all that different, except that we already love our dogs.

One outfitter I met in Wyoming made a principled distinction in his work. He said he would not bait grizzly bears at dumpsters for "rich East Coasters to shoot" if they were not able to find and kill a wild bear. While I wish he'd change his profession, this showed a limit to how much he would manipulate nature solely for someone's enjoyment. Meanwhile, another outfitter said he felt he was "rewilding" his clients, who connected with nature while hunting. Then he laughed and asked me if I would like to see him go out of his sort of "rewilding" business.

Personally? Yes. But I understand that he has to make a living, and I agree that some hunters love nature passionately and work hard to protect it, and not solely because they want to hunt in it. Also, I don't expect everyone to agree with my views. But my hope is that outfitters, hunters, and anyone who harms nature as part of their "enjoyment" of nature will consider lessening those activities and finding alternatives. Hunting and fishing may help some to rewild, but it is not the sort of rewilding I support because I feel we should do all we can not to harm or kill other animals or trespass into and destroy their homes. Rewilding our hearts should help to keep other animals safe, alive, and thriving, and keep intact the various landscapes and homes in which they live.

Fencing the Range: Coexistence and Cattle

Earlier, I mentioned the ecological rewilding program called the Yellowstone to Yukon Conservation Initiative. One of the most high-profile and controversial efforts of Y2Y has involved reintroducing wolves into portions of Wyoming, Idaho, and Montana, where they once predominated. Y2Y did this knowing that more wolves would mean more conflicts with the region's ranchers. In fact, historically, the eradication of wolves was often presented as an economic necessity, and nothing today has changed the issue's basic dynamics. If ranchers gather docile cattle within scalable fences, wolves will help themselves to an easy meal.

I once received a very thoughtful email from a reader on this issue that captured the contradictions and conflicts well:

> As a resident of Montana I've often been confronted with the controversy of hunting wolves. Although it seems like it may be inhumane, how can we justify killing livestock for a burger more so than killing wolves to provide for your family? The majority of Montana is ranchland. There's not a community you can visit that isn't surrounded by

> ranches, filled with livestock, which is basically the sole income of most of the families here. The wolves have been a problem for a long time, especially in this area, because they mercilessly kill livestock, and it can eventually financially destroy any family.... No, we should not go around killing every wolf that lives. But in moderation, it is a great solution. Humans have been hunting animals since the beginning of time for survival. This is simply a matter of such.... The writer of this article... must consider our point of view and how it affects us, not just the urban side of life. With a more educated look at the case, it's plain to see that it's simply a means of survival.

In two significant ways, the writer and I agree. First, I see no moral difference between killing cows for us to eat and killing wolves and other predators so they don't eat the cows before us. Either way, animals suffer. If we did not eat cows — or if we ate them far less often so that there were far fewer, smaller, and better-protected ranches — then this problem would mostly solve itself. However, I also agree that the welfare of the people who run ranches is important. Any ecological effort that tries to preserve wildlife in ways that will inevitably harm people isn't going to be successful. This highlights why ecology, economics, and society often go hand in hand. Protecting wolves while undermining human communities is not compassionate rewilding.

The main ways I differ with the writer are with his defense of more killing as the main solution and with his vilification of wolves as "merciless." Wolves are in fact capable of tremendous caring and mercy, and it's simply presumptive and self-centered to interpret the same act — eating cows to survive — as acceptable for one predator but not for another. Both humans and wolves are primarily concerned with their own survival, and finding humane, nonkilling solutions to resource conflicts often must start by recognizing this. Furthermore, wolves are naturally evolved predators, and they have no choice as to their meal plan. They also are not moral agents and should not be held ethically accountable for their predatory ways.

In this situation, one partial remedy could be to offer ranchers the chance to quit ranching and pursue other sources of income or livelihoods. Given adequate financial support and training, some families might voluntarily change, thus minimizing this conflict. Ranching is not inevitable; it's a choice. Obviously, many ranchers enjoy and prefer ranching, so for them the question becomes: In what ways can cows be protected other than by killing predators?

Several, as it turns out. One inspirational approach has been developed by a Kenyan teen named Richard Turere. Living in lion country and having to protect his family's cattle from being killed by these and other predators, this young Maasai created "lion lights" to keep predators away from livestock. Lion lights consist of outward-facing, flashing LED bulbs powered by an old car battery. As the lights flicker on and off, the lions are fooled into thinking a human is around. Richard is proud that he did this on his own, and in 2013, he was invited to give a very prestigious TED talk about it.

Other examples of ways to scare or ward off potential predators include the use of guard dogs and flagging (fladry) — that is, attaching bright-colored flags to wire fences. Simply cleaning up garbage and animal carcasses around farms and ranches also works, and this is something we've known for more than forty years.

Further, in addition to better protecting livestock, changing people's attitudes toward predators and increasing their tolerance of them is an essential aspect of any effort to find humane solutions. Our feelings about predators vary a great deal and usually depend on a range of social and cultural factors. As mentioned above, sometimes financially reimbursing ranchers for livestock losses is enough to dissuade them from killing predators, but sometimes not. Sometimes more education about the ecological benefits of predators helps, and sometimes the most important factor is perceived social norms. Whether others we know approve or disapprove of killing makes a big difference in our behavior. As conservation biologists Adrian Treves and Jeremy Bruskotter rightly argue, we need much more research to determine the best mix of incentives in various situations, rather than "the haphazard patchwork of trial and error that has in many cases characterized predator conservation efforts."

Biophilic Cities: Rewilding Where We Live

As I've mentioned, human well-being in general is enhanced by exposure to nature. According to cognitive scientist Art Markman, research shows that people who live near parks experience "lower levels of mental distress and higher levels of well-being" than those who do not. So one ongoing effort, which we need to expand, is incorporating more of nature within our suburbs, towns, and cities. In "biophilic cities," urban landscapes and "built environments" are redesigned and rewilded so that they include more natural areas and take practical measures to protect wildlife from human impacts. People are also working hard to rewild much larger areas, such as with the Rewilding Europe Project.

I have had the pleasure of working with Timothy Beatley, Teresa Heinz Professor of Sustainable Communities in the School of Architecture at the University of Virginia, to get animals on the agenda of people responsible for developing and implementing city planning, so that we can expand our urban compassion footprint. Our collaboration is a good example of an unlikely partnership. I first met Tim when he came to Boulder to give a talk in the School of Environmental Design at the University of Colorado, and it was clear that we had strongly overlapping interests. We both wanted to bring animals back into the lives of people who might not have much contact with them because of where they live. We both feel people have an ethical duty to better understand the impacts of community development, design, and planning on animals. Plus, a number of effective strategies and tools already exist for creating and improving human-animal coexistence. Major cities where animals are being nicely incorporated into more-hospitable urban landscapes include Vancouver, Toronto, Amsterdam, and Chicago. I am proud to live in Boulder, Colorado, a truly biophilic city, which has sixty-three urban community parks, neighborhood parks, and pocket parks, along with 263 miles of multiuse hiking and biking paths.

Biophilic cities incorporate a variety of approaches: building green rooftops and green walls, planting native vegetation around homes and buildings, reducing the spatial footprint of buildings, implementing nighttime lights-out campaigns, restricting the use of highly reflective glass and glass facades that disorient birds, enforcing noise restrictions, and developing and maintaining nature corridors where human and nonhuman animals can freely mingle. All these efforts make a significant difference. Toronto has been identified as one of the world's most deadly cities for birds. The Fatal Light Awareness Program, according to the *New York Times*, "estimates that one million to nine million birds die every year from impact with buildings in the Toronto area. The group's founder once single-handedly recovered about 500 dead birds in one morning." Indeed, it's been estimated that 100 million to one billion birds may be killed by collisions with glass windows annually in the United States. We also know that "sky glow" from artificial city lights can disrupt activity cycles of wild animals, and noisy machines can make it difficult for bats to hear the footsteps of their prey, which can lead to nutritional stress.

By taking into account the lives of nonhuman animals within the context of our cities, we can reconnect with many different aspects of nature and rewild our hearts. David Johns, who teaches politics and law at Portland State University, argues convincingly, "There is no substitute for reimmersing people in the world that gave us birth." However, rewilding our cities and neighborhoods can also lead to conflicts. Like the stories I tell that open this chapter, coexisting with animals means accommodating their presence. Sometimes they can be a nuisance to us, and sometimes we might feel afraid if the animal in question is a fellow predator.

For instance, a 2012 news story titled "Wolves and Mountain Lions 'Poised to Invade Densely Populated Cities in the United States'" once caught my eye. As an expert on carnivore behavior, I was shocked by this headline. The story was deliberately alarmist, playing off of human fears to get attention. In itself, this wasn't so surprising; as I discuss in chapter 4, the media often feed our anxieties about animals and misrepresent them in sensational ways. Yet the story's opening claim, that "wolves and mountain lions could soon be a more common sight in densely populated cities," was supported by a number of scientists, wildlife managers, and animal control officers.

This is, simply put, fear-mongering, and it works because people genuinely fear large predators. When it comes to rewilding our cities, we need to be self-aware of our own emotions and projections; we need to take responsibility for them and put them aside. In part, we do that by accepting that coexisting with wild animals can be unpredictable, and so we need to be alert and prepared for potential encounters. We also ease our fears by understanding wild animals better, including their habits and needs. Scientists in particular need to refrain from making exaggerated, alarmist, and unsubstantiated claims that aren't supported by data.

By rewilding our cities and suburbs, will we suddenly find "predators at our doorstep"? No, not in the sensationalist way that phrase is often used — as if animals were just waiting for us to let down our guard and attack. But in fact, the recovery of wild predators within urban environments is a rare and remarkable success story, and if anything, these animals are victims of their own success (see *Nature Wars*, by Jim Sterba). So, as part of rewilding ourselves and our hearts, we need to embrace the animals with whom we share space and time and, occasionally, change our ways to coexist successfully. This is one way we indicate our appreciation for wildlife and actively care about what happens to them and their families and friends. Sometimes, as the stories I tell at the start of the chapter indicate, we need to be wary; sometimes we may be inconvenienced. What is no longer an option is to choose to ignore nature, nor can we continue to blame wild animals for problems that arise from our own ignorance and fears. I have often asked realtors to be up front with people when showing them homes or lots where wild animals live. As Timothy Egan notes in a discussion of the dangers of going out into nature, in this case Grand Teton National Park, "Sometimes the bear gets you." If we don't want our cities to be walled human prisons, we must accept the calculated risks of coexistence — we can build communities that allow space for animals and take reasonable precautions to keep ourselves safe. We can't control everything, but we can acknowledge that animals aren't to blame for behaving as the beings they are.

Models of Rewilding: Everyday Activists and Uncommon Messengers

I fully understand why it is difficult to be optimistic given the numerous and daunting ecological challenges with which we are faced. But there are unexpected victories that help us keep our dreams alive. For example, in April 2011, Bolivia announced it would grant all nature equal rights to humans. Among the eleven rights it names are the right to life and to exist; the right to continue vital cycles and processes free from human alteration; the right to pure water and clean air; the right to balance; the right not to be polluted; and the right not to have their cellular structure modified or genetically altered. In 2012, Colombia banned the illegal trade of night monkeys, and Costa Rica banned sport hunting, the first Latin American country to do so.

I am also continually inspired and motivated by the people I meet when I travel. I travel a lot, and in itself, traveling is no fun. Regardless of the inconveniences, I keep doing it because I always meet people who really care about what's happening in and to the only planet on which we live. These people inspire me to keep going and fighting on behalf of nature and wildlife. I really mean this. No matter where I go, there is always someone, usually more than one person, who is trying to reduce cruelty and to make the world a better place for all beings, human and nonhuman. Some of these people are well-known, dedicated activists, but most are everyday people or folks I would never expect to care like they do, and I'd like to end this chapter by presenting a sampling of them. It's important, really essential, to know that every single individual counts and to realize there are unsung, unheralded heroes all over the world. We need all the people we can to rewild our hearts and to expand our compassion footprint.

For instance, take Howard Wang, who works at the Moon Bear Rescue Centre outside of Chengdu, China, where I regularly visit. Moon Bear rescues bears who have been horrifically abused on farms that serve China's bear-bile industry. To quote Jill Robinson, founder and CEO of Animals Asia, the organization that runs the rescue center:

> Howard has worked here since 2000 and was the very first bear worker we employed. He has expertly risen through the ranks and is now one of our three Senior Bear Managers. He loves the bears and life here with a passion — living on site and throwing himself into every duty — many

> far beyond what he is actually employed to do. He supervises bear teams and houses and helps to look after over 160 bears. When the earthquake struck in May 2008, he joined us on trips to help the Red Cross in their work — initially helping people, to the point where we had gained government trust and were then able to help the dogs and cats. We rescued over one hundred dogs and several cats — and funded their stay in a local animal shelter until the earthquake victims had rebuilt their lives and homes and were able to take their loving family members back.
>
> Howard also said the most beautiful thing one day about the bears — that he was joined to them by a silver thread. I love his smile, too — it radiates love for the bears every day and illustrates just how much people here in China care about protecting and respecting both wild and domestic species.

I agree with everything Jill wrote. Whenever I see Howard, his contagious smile radiates warmth and says everything is working out just fine. We need more Howards.

Once, when I was flying to Chengdu in June 2011 to visit the Moon Bear Rescue Centre, I met a Chinese businessman who, on his own admission, "never spends a second wondering or worrying about the plight of animals and Earth." During our fourteen-hour flight, we had some nice chats, and I think I convinced him that he should care about what we're doing to animals and Earth. Gladly, he told me his children care, and they ask him to be more concerned. He wasn't a "bad guy," but like many people, he was too busy doing other things — like making money to care for his family and developing his business — to worry about the very things that I think and worry about every second of my waking days. However, as we discussed the difference between my vegan airplane meal and his meat-filled one, he agreed that my meal was probably better and better for him. Many others have made the same remark.

Spreading the word to others is important, since many people remain unaware of animal welfare issues, or they feel disconnected to what goes on in other parts of the world. Everyone counts, of course, and I find that many people care, but sometimes they don't know what to do or how to get involved.

In October 2012, I gave a talk about the plight of China's moon bears at the Louisville, Colorado, library, and I was surprised by an elderly couple sitting in the second row. They seemed to be absorbed in every word, and after I was done talking, the woman raised her hand, clearly fighting back tears, and quietly asked if there was any hope for them at all. Yes, there is, I told her. Jill Robinson and her dedicated team are fighting the bear-bile industry with all their heart and soul, and slowly but surely they are making progress. The woman, who hadn't known about farming bears for bile, smiled and said she would become more involved in issues like this, and I gave her the information to do so. Moments like this happen all the time, and it is always so heartening.

I also often receive good news and inspirational emails from people working to make the lives of animals better. On a trip to Pullman, Washington, to speak at a fundraiser for the Whitman County Humane Society, I met many wonderful and dedicated people who, after a long day at the office, work into the night to develop this shelter. I also met some wonderful dogs in Pullman, including Lucky, Yoghurt, Monster Mash, Leon, Baxter, and Brendal. I once received an email from a young girl who wanted the world to become vegetarian and for people to stop going to circuses, and I often get emails from people looking to connect with animal organizations. These positive people willing to work on behalf of animals keep me going. Once, when I was in Almada, Portugal, giving a lecture on animal emotions, I met a woman who was hell-bent on heading to South Africa to go to the frontlines to stop the slaughter of elephants. She was well aware of the risks of such confrontations, but she said she was committed to work to save these magnificent animals.

Wonderful people also visit Boulder, where I live. In February 2012, my friend Louie Psihoyos, the Oscar-winning director of *The Cove*, introduced me to Leilani Münter, a race car driver with a degree in biology who is a formidable, passionate, and tireless animal and environmental activist. Leilani calls herself an "uncommon messenger," and she is able to reach audiences to which most people would not have access. I love her mottos, "Never underestimate a vegetarian hippie chick with a race car" and "Life is short. Race hard. Live green." We need to attract more uncommon messengers who think out of the box and who speak to those caring people outside of the mainstream animal and environmental movements.

Other uncommon messengers include Linda Tucker, who left a successful career in business to become a conservationist and who founded the Global White Lion Protection Trust. Then there is Francisco Mayoral, a poor fisherman in Baja California Sur, who now works hard to protect whales after meeting one face-to-face. And then there's Howard Lyman, a fourth-generation Montana cattle rancher also known as the "mad cowboy," who became a champion of vegetarian and vegan diets.

These and other people are models for what can be done when we are passionate about protecting our magnificent planet.

Today, I hope I can inspire others just as others inspired me when I was young. One of my early inspirations was Barry Commoner, whom I mentioned in the introduction. Another was Michael W. Fox, with whom I worked on my doctorate degree. Like Barry, Michael was way ahead of the times in trying to get people to think deeply about the use and abuse of animals in research and other human venues. In the tradition of classical ethology, he also stressed the importance of watching in detail what animals do when they interact. These two courageous and vocal pioneers made my graduate education the breeding ground for what I have been doing for the past four decades. In many ways, as I reflect back to discussions with these and other forward-looking and "unconventional" scientists, I see the roots of my ideas about redecorating nature, our compassion footprint, and rewilding our hearts. We never truly know what seeds we plant, or how they will flourish, but with patience and care, some will. These ideas were already existing in my brain and in my heart way back then. In fact, my parents told me they were there when I was a mere three years old.

4. Rewilding the Media

Our Mirror Up to Nature

Since June 2009, I have written regular online essays for *Psychology Today*. When I agreed to do this, one of my main goals was to become something of a media watchdog concerning how nonhuman animals are portrayed in the media and popular culture. To me, this is a hugely important issue. Animals in the media are often treated as mere objects; they are sensationalized, and they are misleadingly "humanified" to suit our own ends or beliefs. We should expect, and insist, that animals in the media be represented accurately as the beings they are, not as who we want or imagine them to be. Accurate reporting is part of the process of rewilding, as is becoming self-aware of how images of animals, and even the language we use, both reflect and influence how we feel about other animals — and thus how we treat them.

The media are an extremely powerful influence, and the messages they convey make a tangible difference in the actions we take. For instance, one study has shown that accurate information about climate change is "the best predictor of an intention to do something about it." Furthermore, the study's authors note that emotion is a very important predictor of sustainable behavior. The same surely holds true in our interactions with other animals. Eleonora Gullone (in her 2012 book *Animal Cruelty, Antisocial Behavior, and Aggression*) argues that the media's increased attention to animal cruelty has led to an increased acceptance of the link between human antisocial behavior and animal abuse.

By looking at the media, we clearly see the "human dimension" in our relationships with nature and nonhuman animals. Movies and documentaries, the bias in our language and news reports — all these things reflect how we understand the world and what we believe. Many times, pop culture myths and sensationalism can influence us strongly without our even realizing it. As we rewild our hearts, we have to take personal responsibility for these issues. We have to examine our beliefs and the way we talk, and we should question media portrayals and not take them at face value. Far too often, they are misleading, sometimes dangerously so. Often this is because "sensationalism sells," experts on both sides of an issue haven't been consulted, or the issues at hand are too complex to be handled in a short radio or TV report.

The question of how to accurately portray animals is not new. As Ralph Lutts, in his outstanding book *The Nature Fakers*, notes, in many ways the science versus sentiment debates that arose in the early 1900s still influence our thinking. Around the turn of the twentieth century, nature writing became extremely popular, as were questions about the nature of animal minds, whether nature was "red in tooth and claw" or more peaceful, and whether more subjective, individualized portraits of other animals were accurate and scientific. Teddy Roosevelt and John Burroughs attacked many popular nature writers, including Ernest Thompson Seton, Jack London, and William J. Long, as "sham naturalists" who sentimentalized and misrepresented animals in popular media. This bias to treat animals as interchangeable objects, rather than individual subjects, persists today. And I am happy that Rupert Sheldrake and I were responsible for having Long's book *How Animals Talk* reprinted in 2005, as it is an outstanding work on animal behavior and animal minds. I often wish I lived back then and had been Long's student, accompanying him and others on their forays into nature.

In contrast to dismissively objectifying animals, the media also frequently sensationalize animals in overly dramatic, fear-inducing ways. Because "blood sells," stories of violence are often exaggerated to get attention, which leads an unknowing public to fear the "predators at our doorstep" (as I mention in chapter 3). Perhaps sensationalism will always occur, especially in movies, since moving people emotionally is how studios sell tickets and how news organizations get people to click on websites and view pages. But we can still learn to recognize and question inaccurate portrayals when we see them.

However, I've also learned a few things by joining the popular media through my *Psychology Today* column: First, judging by the reactions to my columns, people feel passionately about animals and about the complex and paradoxical nature of the human-animal bond. Further, the interactivity of the Internet means that people no longer must passively accept whatever they read. They can respond to authors and the media directly, and they do — sometimes only to shout, but more often readers want to thoughtfully engage the ideas and proposals. And last, people hold an incredibly diverse range of opinions, and the exchanges I have witnessed and been part of about human attitudes, beliefs, and values regarding nature have been eye-opening and invigorating. Even when I do not always agree with someone's perspective, I always learn from these exchanges, and I hope my readers feel the same way.

What We Say: Rewilding Language

The words we use to refer to animals strongly influence how we view them and the actions we take to protect them. They can also be very revealing, particularly the pronouns we use. With our companion animals, people typically use gender-specific references: their dog is "he" or "she," not "it." This is because we feel close to our dog or cat, and our language honors the animal's subjectivity and reflects our emotional connection.

But words can also be distancing mechanisms. When it comes to the animals humans eat for food, we often use impersonal language, as if to avoid presenting the animals as thinking, feeling subjects. An animal will be referred to as "it" and "that" rather than as "he" or "she" or "who." As Georgia State University's Carrie Packwood Freeman points out in her essay "This Little Piggy Went to Press," a cow on a plate is called meat or a hamburger, and a pig is called bacon, sausage, or ham. In other words, people will ask for a bacon, lettuce, and tomato sandwich, but does anyone ever order a Babe, lettuce, and tomato sandwich?

So, we need to be careful and self-aware about the words we use to refer to other animals. When I talk about the food people choose to eat, I point out that very often it's a matter of *who's* for dinner, not *what's* for dinner, because the animals who wind up on our dinner plates were once alive and sentient. They are not objects, and our words should not objectify them. This small change in words has resulted in a number of people deciding to change to vegetarian or vegan diets. In June 2012, I received a most heartwarming email from a woman in Vienna, Austria, who had heard me speak, and she said, "You'd turned at least four people to vegetarians — including myself."

In the end, perhaps one of the simplest and easiest ways to rewild our hearts is to rewild our language.

Animals in the Movies: Myths and Make Believe

Perhaps it should be obvious, but I often find it isn't: Movies aren't real, and more often than not, movies do not portray animals accurately. Talking dogs? Singing warthogs? These characters are obviously human stand-ins. They are not to be taken seriously. Walt Disney, of course, was a master at anthropomorphizing animals and using them to re-create fairy tales or to embody human myths about the wild. And yet, as much as we might understand that an animated fantasy isn't realistic, these movies can have an enormous influence in how real animals are perceived. A few excellent books about this include *Reel Nature* by Gregg Mitman, *Animals in Film* by Jonathan Burt, and *Shooting in the Wild* by Chris Palmer. In short, entertainment has direct consequences for animals.

Live-action movies, ones that use live animals and are set in the real world, are perhaps the most insidious, since it's easier to forget that what we are seeing is a human projection. But animation influences us as well. In movies, animals are often portrayed according to their cultural stereotype, which only reinforces those stereotypes, even if these ideas bear no relation to the actual animals. Are lions noble and hyenas venal? Are mice cute and loving and rats dirty thieves? Are snakes and sharks evil, mongooses brave, chickens stupid, and foxes clever? Folk tales and movies lead us to think so.

Hollywood and TV have a widespread effect on how animals are perceived, and for the most part, they don't care whether animals are presented realistically. They only care about creating drama, and animals are often staged and their behavior manipulated to fulfill a story's needs. For instance, the 2011 movie *The Grey* presented a pack of violent rogue wolves who attack the film's human protagonists. In a way, this was only a hyperviolent retelling of "Little Red Riding Hood," one that played off of the negative myths of grey wolves as vicious, human-hunting predators. Nothing could be further from the truth. There have been only two fatal wolf attacks on humans documented in North America. If *The Grey* were a documentary, it would be guilty of fear-mongering sensationalism. As a movie, it can shrug this off and freely make wolves the villain. Hey, it's just fantasy, right? Yet it is also a pack of lies. It rivals *Jaws* in its misconceptions about wildlife.

A movie like *The Grey* feeds our fears about the real animal. Despite objections by scientists, the US Fish and Wildlife Service has put forth a proposal to remove grey wolves from the list of threatened and endangered species under the Endangered Species Act, so that they would no longer be protected, and in some places wolves are wantonly killed by ranchers to protect their livestock (see chapter 3). If we believe (contrary to what is known) that wolves mean to do us harm and will kill humans if given the chance, then killing wolves becomes a justifiable form of "self-defense," and the idea of coexistence with these magnificent, social mammals seems foolish and unrealistic. A movie like *The Grey* provides the perfect motivation and rationale for "getting rid of" wolves, rather than protecting them.

This is only one example among many, but it embodies a particularly cruel irony. As we know, long ago, the dog evolved from the wolf, and today we consider the dog "man's best friend." But wolves must have been our "best friend" first, or else some wouldn't have domesticated into human-loving dogs in the first place. That original affinity between wolf and human — that enduring friendly cooperative impulse that led to the first instance of domestication we know about — remains, and it's a story the movies rarely tell.

Of course, films sometimes do "get it right," and the power of storytelling can be used to aide rewilding. I remember a great moment in the 2010 movie *The Switch*, in which Sebastian, a young kid, refuses to eat duck at dinner and won't blow out candles on his birthday cake until someone offers to adopt a three-legged dog at a kill shelter. Sebastian also comments that we move too rapidly and that's why we are called the human race! I was thrilled to see this scene, which perfectly mixes humor with a serious message: that there are far too many unwanted dogs. If they chose to, people making movies could go a long way to encourage their audience to reflect positively and proactively on our interactions with other animals.

No Animals Were Harmed in Making This Film

Movies can also mislead us in a more deliberately self-serving way. The incredibly talented and highly trained animal actors who befriend humans, save the day, or otherwise play roles in the drama are sometimes horribly abused, despite claims by movie studios to the contrary. Indeed, there is solid evidence that animal training for entertainment really involves "breaking" the animal to get them to do what's needed, whether they are performing in circuses, zoos, or films. A recent example concerns Tai, the female elephant star in the 2011 movie *Water for Elephants*. Tai became an "emotional wreck" because of the abuse to which she was subjected in the making of this movie.

Another example is the 2012 film *The Hobbit: An Unexpected Journey*. While no animals were harmed during the actual filming itself, according the American Humane Association, it turns out that perhaps twenty-seven animals were killed, and more injured, where they were being housed. The list of maimed or killed animals included five horses, a pony, several goats, sheep, and chickens, and a studio spokesperson for the film admitted, "The deaths of two horses were avoidable." Animals fell into sinkholes, were injured by wire fencing, and were harmed or killed by other animals in the overcrowded conditions.

While animal abuse during filmmaking is sometimes deliberate, more often it is not the intention of the people using the animals. Yet it occurs all too easily. In itself, a movie set can be unsettling to an animal. They do not like the bright lights and noise of a typical set. Even when their basic needs are taken care of, most would no doubt prefer to be left alone.

One potential solution to this problem is exemplified by the incredibly popular movies *Rise of the Planet of the Apes*, *Avatar*, and *Noah*. Only computer-generated nonhuman animals were shown in these films, and these movies are great examples of how it's now possible not to use any real animals to make films. *Rise of the Planet of the Apes* is particularly stunning in this regard, for Caesar, the main animal character (played by Andy Serkis), looks and acts like a real chimpanzee. It's wonderful that not a single chimpanzee had to be used to show viewers what it's like to be a chimpanzee. Caesar's emotional states are beautifully portrayed, and at times I caught myself wondering what this chimpanzee must have felt about being used in a movie this way, only to remind myself that Caesar wasn't a real chimpanzee. While we should continue to do more to protect animals in entertainment, this movie can be a model for ending the use of animals in film.

Reality TV: Nature Documentaries and Photography

One place where we expect accurate portrayals of animals is in nature documentaries and nature photography. After all, that's what they are selling: real life and authentic science. But nature shows can sometimes be no more realistic than reality TV programs like *The Bachelor*.

At their worst, nature documentaries amount to "nature porn." They concentrate on attention-getting sex and violence, showing animals mating, being aggressive, and in acts of predation. The details may be correct, but the impression radically misrepresents animal lives and feeds the false notion that nature is a competitive, brutal game of "survival of the fittest." In fact, wild animals spend much of the time resting and being nice to one another, playing, sharing food, and cooperatively defending food and territory. Indeed, for all species that have been studied, more than 90 percent of their time is spent performing in positive and friendly ways, or what are called prosocial behaviors. Conversely, nature programs will get laughs by showing animals making mistakes and falling down, but animals are not as daffy as they are sometimes made out to be.

Many professionals are also concerned with the accuracy of nature photography. Sometimes, these shots are staged at game farms, where individual wild animals spend most of their lives in tiny, filthy cages, with no regulations governing how they are housed. The "perfect" photos are then sold at a high price to an unaware public, who do not realize they're actually buying an illusion of a wild animal. This raises many ethical questions, since these animals are essentially "paroled and paraded for profit," according to world-renowned "Images of Nature" wildlife photographer Thomas Mangelsen (who kindly provided the photo on the cover of this book). Mangelsen notes:

> Not only am I concerned about the welfare and exploitation of the animals, but also about the continuing loss of credibility and integrity to the wildlife photography

> profession once people learn that many of the photographs they have admired are of animals that spend their lives in cages. I have heard all the rationales for photographing at Game Farms. I find most of the justifications hard to accept and feel most are self-serving and generally don't really consider the welfare or the best interest of the animal or for that matter what's best for the profession of wildlife photography.

Of course, this does not mean that *all* documentaries and photographs of animals are misleadingly false. But it does mean that we shouldn't take every claim of authenticity at face value, and we should remember that entertainment is meant, first and foremost, to be entertaining. Just like reality TV, events can be manipulated to enhance the drama.

There are excellent organizations teaching and encouraging responsible filmmaking, like the Center for Environmental Filmmaking in Washington, DC, as well as an increasing number of moving films that "show it like it is." These include the 2013 documentaries *Blackfish* and *The Ghosts in Our Machine*. *Blackfish* documents the wanton and continued abuse of killer whales (orcas) in captivity, focusing on the life of Tilikum, a male who killed two people and was used as a "breeding machine" while in captivity, his latest home being SeaWorld in Orlando, Florida. This film was very influential: It resulted in schools canceling class trips to visit SeaWorld, and it spread the word that killer whales (and other aquatic animals) really are driven crazy by swimming in circles and that they are often abused while being trained to perform stupid and unnatural tricks solely to entertain humans and make lots of money.

The Ghosts in Our Machine, directed by Liz Marshall, is an incredible and forward-looking film. It follows internationally renowned photographer Jo-Anne McArthur over the course of a year as she documents the stories of individual nonhuman animals who are caught in the web of so-called "civilized society" in the United States, Canada, and Europe. She looks at just those sentient beings whose sentience has been rendered invisible to us, and so they are the "ghosts" in our societal engine. In a promotional clip for her film, McArthur says she feels like a "war photographer" because we really are waging war against these animals as we wantonly exploit them in myriad ways. This is the kind of film that can rewild our hearts by waking us up to the ways we alienate ourselves from the nonhuman animals in our lives. It also stresses that animals are not ghosts — invisible objects — but rather sentient beings who care very much what happens to them. Spend five minutes with Jo-Anne and you can feel the passion that drives her stellar work.

Animals in the News

People always ask me if I've seen or read this or that news article or book about the lives of nonhuman animals, and I often feel a need to correct what I find. The news media often take liberal license to stretch the facts or to simply misrepresent other animals. One of my colleagues, a bestselling author, once told me that it is okay for him to take liberty with facts because he is not a scientist. I bristled and refused to write endorsements for his books.

This is an unacceptable attitude. The long-established ethics of journalism shouldn't be suspended for animals. Clearly, news media leave an enormous footprint in the human cultural landscape, and there are hidden costs and collateral damage when animals are misrepresented, disparaged, and objectified. Journalists "script" nature to fit the different formats of magazines, newspapers, radio, and television, but rewilding demands accuracy in reporting. News media play an important role as we rewild, for the vast majority of people rely on media to learn about the fascinating lives of other animals as well as about the damage we cause to them and our planet.

One issue is that news reporters, particularly on TV and radio, often present animal stories in entertaining ways that make light of animal suffering. For instance, in October 2012, a video of chimpanzees tormenting and tossing around a raccoon at the St. Louis Zoo was shown on a local CBS affiliate and treated as a source of humor. After one news anchor laughs about the "monkey business" in the clip, both anchors giggle while the tormented raccoon tries desperately to escape from the chimpanzees, who keep grabbing at the raccoon until he or she disappears in a drain pipe. Still laughing, one anchor ends the segment by saying, "I've wanted to do that to some of the raccoons in my backyard."

I have a good sense of humor, but there was nothing funny in the chimpanzees' behavior or the raccoon's stress. If it had gone on longer, the chimps' excitement could easily have escalated into a full-blown and deadly attack. Most of all, though, the anchors acted cavalierly and indifferently. If the raccoon were a dog or a cat, my guess is that the anchors would have exhibited heartfelt concern, although raccoons are no less sentient than dogs or other mammals.

Broadcasters have a high profile, and they should try to be clear, positive examples of respect, compassion, and empathy for all animals, human and nonhuman. That's not really asking too much. Fear and suffering should not be fodder for light entertainment and the object of ridicule.

A similar thing happened in a 2009 National Public Radio (NPR) report summarizing a study showing that ants seem to be able to count. An ingenious experiment showed that ants are able to count the steps they take using what are called "pedometer-like" cells in their brain. Fascinating, indeed. Yet part of the experiment entailed cutting off parts of the legs of some ants, and in the NPR report, this was described as a "makeover." This was, I imagine, a supposedly tongue-in-cheek way to make light of the uncomfortable fact that these insects were mutilated to prove how smart they are. Lopping off an essential body part should be unethical in any research, even on ants, and calling it a "makeover" really upset me, since it encourages listeners to ignore the callousness of this experiment.

Finally, it's important to remember that, particularly when it comes to issues of animal welfare in our society, we often don't get the full story. The news media often distort facts or miss them entirely, either because of an inherent indifference to animal issues or because of the influence of powerful business and political interests. In fact, many people working with and for animals are afraid to speak the truth to media or are told not to do so. This adds insult to injury and compounds the problem of the misrepresentation of animals in media. Carter Niemeyer, author of *Wolfer*, once worked for Wildlife Services as a "hit man," someone who kills predators that menace domestic wildlife. In a 2011 interview, Niemeyer said, "Without a doubt, wildlife biologists, who are professionally trained, are inhibited from speaking and acting on their knowledge about wildlife management and resource conflict issues, mainly due to politics within their states." Concerning myths that are thrown around by antiwolf people concerning wolf predation on livestock, Niemeyer said, "I think what's going on is a clash of cultures. The truth as I see it, is that livestock losses attributed today to wolves and other predators are being exaggerated because of this clash.... I never bought into the belief that wolves are wiping out the deer, elk, and moose in the Northern Rockies. Overall, elk are doing great in Idaho, Montana, and Wyoming and are at or above management objectives."

Media Misrepresentations Have Real Consequences

I was talking with a friend as I was writing this chapter and told him how I hate seeing animals dressed up as humans. This misrepresents animals, and even when they are presented as "cute," there are hidden costs. A few weeks before, a *Psychology Today* cover story entitled "Are You with the Right Mate?" was illustrated by a photo of a woman holding hands with a tuxedo-wearing male chimpanzee. Other pictures with the article included a dressed chimpanzee riding on the back of a scooter. The chimpanzee had absolutely nothing to do with the topic, and all I can guess is that the great ape represented the difficulties women have finding acceptable male companions. Cute; nothing more than a cheap visual joke.

But it really upset me because chimpanzees are endangered. My friend understood what I meant but felt I must be overreacting. After all, he said, if chimpanzees are being dressed like humans on the cover of *Psychology Today*, it suggests there are plenty of them to go around, and their endangered status must not be that dire. This, I told him, is exactly the problem. Studies show that misrepresentation in the mass media actually harms their conservation status; if we see a lot of chimpanzees on TV, we believe there must be a lot of them in the wild. Even the Lincoln Park Zoo in Chicago, Illinois, asked for a company to pull their 2012 Super Bowl commercial showing a chimpanzee because this desensitizes viewers and sends the wrong message that these great apes are not endangered and even make a good pet. They don't!

One 2011 study by Stephen Ross and his colleagues shows "the public is less likely to think that chimpanzees are endangered compared to other great apes, and that this is likely the result of media misportrayals in movies, television, and advertisements." Their detailed study demonstrated "that those viewing a photograph of a chimpanzee with a human standing nearby were 35.5 percent more likely to consider wild populations to be stable/healthy compared to those seeing the exact same picture without a human. Likewise, the presence of a human in the photograph increases the likelihood that they consider chimpanzees as appealing as a pet. We also found that respondents seeing images in which chimpanzees are shown in typically human settings (such as an office space) were more likely to perceive wild populations as being stable and healthy compared to those seeing chimpanzees in other contexts."

Another 2011 study by Kara Schroepfer and her colleagues also discovered that the use of chimpanzees in commercials negatively distorts the public perception of chimpanzees regarding their conservation status and that this distortion can hinder conservation efforts. Obviously, these are unintended consequences, but the impacts are real. We need to be aware of the subtle messages we internalize from media and film portrayals of animals. Even positive characterizations can lead to negative impacts for animals if we aren't careful.

The 2003 animated film *Finding Nemo* was a huge success, and it was even a caring (if fictional) representation of fish. Yet about five years later, it was reported that clownfish populations had fallen by around 75 percent in some areas because children wanted their own Nemo as a pet.

In short, part of rewilding our hearts is becoming self-aware of our own sometimes unconscious attitudes toward nonhuman animals and nature. We frequently dress the world and animals in our own image, or believe what's most comfortable, but those beliefs don't necessarily reflect the reality and circumstances of other animals. We must learn to unmask our hurtful assumptions and to question the media's inaccuracies and misrepresentations. Of course, it is also important to recognize when media accurately reflects the lives of the other animals with whom we share our planet. For example, we can freely applaud films that use computer-generated images of animals rather than live animal actors, and we can take heart from documentaries and movies that accurately portray wildlife and the plight of captive animals in society. The media leave a huge footprint in our culture, and animals will benefit or suffer depending on how the media handle this responsibility. The best way to assess media is to consider things from nature's perspective and to honor all beings and their homes.

5. Rewilding the Future

Wild Play and Humane Education

If you don't know how to fix it, please stop breaking it.... You are what you do, not what you say. What you do makes me cry at night.... Please make your actions reflect your words.

— Severn Suzuki

When she was nine years old, Severn Suzuki started the Environmental Children's Organization (ECO), and when Severn was twelve, she and two other young members of ECO paid their own way to attend the United Nation's Earth Summit in Rio de Janeiro in 1992. While there, Severn gave a speech at the close of a plenary session, in which she asked the attendees: "Who do we think we are?" As the quote that opens this chapter shows, she challenged world leaders to do more than talk and debate. She urged them to act, and act now, and act in ways that we know will help animals and ecosystems, not continue to destroy them.

Whenever I watch this speech, tears of joy and deep concern come to my eyes. And I feel a huge surge of inspiration. How lucky we are to have passionate youngsters caring about the future of our planet. Youngsters are card-carrying "biophiliacs" who have an inherent love of life. They are intuitive, curious naturalists and sponges for knowledge. They are also our future leaders on whose spirit and goodwill we and the entire world will depend. It is vital that we teach children well — that we infuse their education with kindness and compassion and a deep immersion in nature, so that they do not grow up "unwild" and so their decisions are founded on a deeply rooted, reflexive caring ethic.

Severn Suzuki was right on the mark. She had reason to be upset and angry. For too long we have acted with no concern for future generations, who will inherit the innumerable environmental messes we leave behind. As Jane Goodall puts it, we are stealing the future of our children by wantonly decimating our planet today. Further, our education system itself needs to be rewilded, since the cumulative effects of technology, media, and our cultural alienation from nature negatively impact children just as much as nonhuman animals.

For instance, in March 2013, Valerie Belt, an animal activist and a grade-school teacher in Los Angeles, sent me an email about an experience she had in the classroom. She gave her students subtraction story problems that used animals as examples, and she asked the students to draw a picture to illustrate each problem. "For example," she wrote, "rabbits on the grass in the forest, lobsters on the ocean floor, or killer whales swimming in a pod in the ocean. When I said this, one kinder student commented about the killer whales in the ocean: 'I thought you could only find them at SeaWorld.' "

This story startled and profoundly disturbed me. It drove home that we can't assume children automatically know even basic facts about nature and animals. We have to teach them, and we have to be role models. We have to create a world in which humane care and awareness of nature are givens. In fact, it shouldn't be surprising at all that if youngsters don't spend time in nature, and aren't taught about the natural world, that they will suffer and remain ignorant. This chapter, then, looks at the negative impact of unwilding on children, at how to improve humane education, and at the need for more research in conservation psychology, so we understand our relationship to nature better.

Kids are a deep source of energy and inspiration, and I work with them every chance I get, such as with Jane Goodall's global Roots & Shoots program. Clearly, youngsters connect easily with animals, and the compassion they show gives me hope for the future. In March 2011, I gave a talk in Idaho Springs, Colorado, about the emotional lives of animals. After my talk, there were many questions from both kids and adults about the fascinating lives of animals and deep concern about how we mistreat so many of them for our own ends. Just as I was wrapping things up and saying, "Thank you for coming," a small boy in the back started waved his hand wildly. People were getting up to leave, but I asked them to stay for a minute so that I could answer his question. The boy stood, said his name was Luke, and very proudly and boldly asked: "Why can't we just leave them alone?" Just about everyone in the audience applauded his courage to stand up and be heard. Afterward, I walked out with him and his parents, and I was really impressed with his commitment to save animals. Luke's question summarized what it took me a whole book to articulate in *The Animal Manifesto* — namely, that if animals could talk, they would say to us, "Treat us better or leave us alone." Kids get rewilding naturally, but it's up to us to rewild the world in which they live and learn.

When Kids Unwild: Animal-Deficit Disorder

In 2005, Richard Louv coined the term "nature-deficit disorder" in his book *Last Child in the Woods*. Since then, the idea has caught on in both scientific and popular media. To me, nature-deficit disorder clearly includes what I call animal-deficit disorder. When youngsters and adults fail to get out into nature, they not only lose contact with landscapes and nature but also with the animals who live there.

Not everyone likes these terms, but I do not know anyone who would disagree that kids need to get off their butts and play outside more for their own physical and mental health. Numerous researchers and psychologists have traced how our childhood interactions with nature and animals influence our development, in positive and negative ways; for instance, see *The Human Relationship with Nature* by Peter Kahn Jr. *Why the Wild Things Are* by Gail Melson, and *The Significance of Children and Animals* by O. E. (Gene) Myers. Today, it's a plain fact that kids experience nature less and less. In 2010 it was reported that kids in the United States spend an average of eight hours a day watching television, surfing the Internet, playing video games, and consuming media, and few parents place any rules on how much time their children spend on these activities. In March 2012, the BBC reported that children in the United Kingdom are "losing contact with nature at a 'dramatic' rate, and their health and education are suffering." The renowned Swiss naturalist and educator Louis Agassiz once famously said: "Study nature, not books." I agree.

Childhood has changed the world over, becoming more sedentary and less wild, and not just because of the rise of media and communications technology. In July 2011, I had the pleasure of attending an incredible international meeting in Wales called "Playing Into the Future — Surviving and Thriving." Around 450 delegates from fifty-five nations spoke on the importance of play and why it's decreasing or sometimes altogether missing in children's lives. One point I made is that today there simply are far too many of us; higher population densities lead to reduced play among nonhuman animals, particularly where resources like food and shelter are limited, and the same dynamic occurs with us. Others noted that seriously ill children often don't play because they can't or because mothers or caregivers may stop them to conserve their energy. Many parents and families can't afford time to play because the children must work, or families curtail play because there aren't any convenient, safe places outside the home to go. These issues obviously affect poor neighborhoods and countries, but affluent neighborhoods and areas also suffer from less free play outdoors.

A recent survey of children's books provided another indication of this ongoing "unwilding" trend. After looking at 8,100 images in 296 award-winning children's books published between 1938 and 2008, researchers charted a clear decline in the representation of nature and animals. Over this seventy-year period, it was found that the primary environments changed dramatically. Built environments grew from about a third to over half of the images in that period, while natural environments declined from about 40 percent of the images to roughly 25 percent. This same decline was true for images of wild animals and domestic animals, leading one survey author, Al Williams of the University of Nebraska-Lincoln, to write: "The natural environment and wild animals have all but disappeared in these books."

All of these findings are bad news in many ways. As I noted earlier, unwilding translates into less concern about the environment and a lack of connection with nature. Whether this amounts to a bona fide "disorder" is debatable. As many have pointed out, the term "nature-deficit disorder" is descriptive, as are my terms "unwilding" and "rewilding"; they are metaphors, not a diagnosis or accepted medical condition. As Wendy Russell, a senior lecturer in play at the University of Gloucestershire, notes, the interrelated issues of nature, play, technology, and health are complex. She writes:

> The binary opposition of nature and technology, outdoor and indoor, is a false one and is leading to an oversimplification of the issues at stake. This was summed up beautifully by one of our postgrad students posting online a photo of his teenage son and friends, up a tree, on their iPhones. Online, for some children and young people, is the only space they can find that is relatively adult-free; perhaps the message from this is less that we should be taking children outside to do adult-supervised worthy

> play activities and more that we should think seriously about how the public realm is organised in such a way that children won't or can't play there without adult supervision. We need to stop trying to organise children's play and start thinking about the conditions that support it.

In a similar vein, Joe Zammit-Lucia wrote to me: "I am also concerned by the construction of terms such as 'nature-deficit disorder,' as this implies some sort of ailment/maladjustment. Is this really the case or is it just a function of a changing world? Is it merely the equivalent of saying that I have paper-and-pencil-deficit disorder because I do all my writing on a computer? A friend of mine relates a story where a father took his teenage daughter on a long and 'wonderful' hike ending in a spectacular waterfall. On arrival, the daughter's reaction was, 'Dad, is this why you made me walk all this way? If I wanted to see a f***ing waterfall, I could just have gone to the mall!'"

There are many ways to react to this story. One is to respect and honor the resilience of children. They can adapt and thrive despite immense challenges and despite our worries that they won't. Many seem to find their own way, and it may be different from previous generations'. I guess that's what they mean by a generation gap. But the story also illustrates just the sort of disconnection from nature rewilding seeks to heal. Nature is not a fad that goes out of fashion, nor is it an "old technology" that has or will become obsolete, the way rotary phones and typewriters are now museum pieces. In fact, nature is the original; it's what mall fountains are imitating.

If children lack an appreciation for being in nature, if falling water is all the same no matter where it happens, then that represents our collective failure as a culture and a society. We need to figure out how to foster a love of nature and other animals so that every generation sees this connection as precious and vital and worth nurturing. We must be inspirational role models. There is no doubt that kids do indeed suffer when they don't spend time outdoors and interact with other animals. Unwilding affects them. As Richard Louv notes, "Nature experience isn't a panacea, but it does help children and the rest of us on many levels of health and cognition. I believe that as parents learn more about the disconnect, they'll want to seek more of that experience for their children, including the joy and wonder that nature has traditionally contributed to children's literature."

Wild Play

Better a broken bone than a broken spirit.

— Play Wales

There are many reasons why children need to play, just as young animals need to play. We need free-ranging kids. They must be allowed to get dirty, to learn to take risks, and to negotiate social relationships that might be complicated, unexpected, and unpredictable. I love the slogan of Play Wales, a Welsh charity promoting children's play, "Better a broken bone than a broken spirit," attributed to Lady Allen of Hurtwood. We should all embrace it with all our heart.

And we should all embrace the notions of "wild play" and "green play." Can we ever play too much? I don't think so, or at least I don't see it as a significant worry. Play embodies rewilding on so many levels. It's the simplest and most natural thing to do.

Mark Hoelterhoff, a psychologist at the University of Cumbria, has written about the importance of "green play" for mental health. He notes that playing outside reduces the symptoms of ADHD better than playing indoors and that playing outside is good for developing self-discipline. Hoelterhoff states it concisely and correctly, "Nature and children are natural playmates — they're both wild and messy, unpredictable and beautiful. Kids just don't play out like they used to." Psychologist Susan Linn notes, "Time in green space is essential to children's mental and physical health.... And the health of the planet depends on a generation of children who love and respect the natural world enough to protect it from abuse and degradation."

Further, a study published in *Conservation* in 2012 showed that "spending time in nature without computers, phones, and other electronic devices makes people more creative." It is still not known if the "cognitive advantage" or "creative bump" was due to increased exposure to nature, decreased use of technology, or other factors. However, because an increased exposure to nature and decreased use of technology usually go hand in hand, the authors write, "They may be considered to be different sides of the same coin."

Separate from developing a connection to nature, social play is extremely important in and of itself. Since I have studied social play behavior in various animals, particularly dogs, wolves, and coyotes, teachers and child psychologists occasionally ask me, "What can we learn from the way in which animals play that will help us gain a better understanding of human play?" Spontaneous social play is indeed critical, and many people are rightfully concerned that it is happening less and less among today's youngsters. The study of play behavior in animals tells us a lot about what human children need. Basically, the range of benefits includes social development and socialization, physical exercise, cognitive development, and also learning the social skills of fairness, cooperation, and moral behavior (or what my colleague Jessica Pierce and I have called "wild justice"). Play behavior in animals mimics aggression and competition, but in fact play is defined by displays of cooperation, empathy, and compassion. For example, the basic rules for fair play in animals are as follows: *ask first*, *be honest*, *follow the rules*, and *admit you're wrong*. When the rules of play are violated between animals, and when fairness breaks down, so too does play.

Obviously, this dynamic applies to humans. Kids learn these same rules and lessons when they play. In fact, play researcher and scholar Bob Hughes feels that play has had an evolutionary benefit. Hughes coined the term *evolutionary playwork* (in his book of the same name) "to reemphasise that the growing body of scientific evidence confirming a direct relationship between play, evolution, and brain growth, demonstrated that playwork should never have been viewed either as a social engineering, a socialising or citizenship tool, but rather as comprehensive support for deep biological processes — expressed through mechanisms like adaptation, flexibility, calibration, and the different play types — that enabled the human organism to withstand the pressures of extinction." Thus, Hughes writes, "playwork was about helping the species to survive extinction and adapt to change, by ensuring that wild adult-free play in diverse environments was still a choice for its children."

Play is essential for the psychological well-being of the child, and not just any type of play, but social activities that are unscheduled, unplanned, and unmanaged by adults. As an evolutionary biologist, I see Hughes arguing that play is vital to thriving, surviving, and reproducing. In evaluating the effectiveness of various types of play, Hughes lists what he calls "bio-outcomes." Bio-outcomes include "an increase in brain size and organisation, an increased ability to roll with the punches, improvements in resilience and optimism, greater mental flexibility in problem solving, the development of cortical maps, and an increase in successful adaptive strategies."

Hughes is very clear about the type of play that doesn't work: "If the activity is bounded by adult rules, if it is stiff, formalised, and dominated by the need to score points and flatter one's ego; that is not play, it is something else." Play contains risk. Echoing Play Wales, Hughes notes, "Play, like life, is not safe, and if it is, it is not play," which is why "a broken arm now might save a life later." In my studies of animals, I always like to say that play is fun but it's also very serious business.

I agree with Hughes, and I love the way he puts the issues. Practicing with "small" risks is how we learn and polish our survival skills. Along these lines, I would add that play is training for the unexpected. In earlier research, my colleagues Marek Spinka, Ruth Newberry, and I have proposed that play functions to increase the versatility of movements and the ability to recover from sudden shocks, such as the loss of balance and falling over. It enhances the ability of animals to cope emotionally with unexpected stressful situations. To "train for the unexpected," animals actively seek and create unanticipated situations in play and actively put themselves into disadvantageous positions and situations.

Near the end of his book, Hughes writes, "So my wild playing child, as a representation of everything human that has gone before, is as ancient and ageless as the land; the wise sage and the awestruck newcomer; the timeless survivor and the passionate explorer." Oh how true this is! Unsupervised social play in nature is not just what the doctor ordered, it's an expression of our evolutionary heritage in which we develop the very attributes of empathy and compassion we most need today.

Rewilding the Classroom: Humane Education

A significant part of rewilding our society involves focusing on future generations and rewilding education. How we educate children about nature, nonhuman animals, and our shared home needs to change. As my colleague A. G. Rud says, we may need to "unschool" youngsters, since too often "school" means sitting at a desk indoors, reading and repeating facts, taking written tests, and worrying about grades and status. Rewilding education means learning about nature firsthand — while outside — and it would involve meaningful play, in which who you are and what you do are as important as what you know.

Today, much of this falls under the umbrella of what's called "humane education," which often incorporates conservation education. Humane education focuses on teaching moral intelligence and reverence for all life, following the significant contributions and suggestions of Albert Schweitzer. It strongly encourages coexistence, compassion, and peaceful relationships among all beings. It also focuses on the learning environment and seeks to foster curiosity over rote learning, as well as social intelligence and respect for others. I would add that rewilding education would also mean a greater focus on animal studies and, in the field of anthrozoology, the study of human-animal relationships.

In 1996, Howard Gardner, the champion of the notion of multiple intelligences, added "naturalistic intelligence" to his original list of seven intelligences as one of the main types of intelligence all people, to one degree or another, possess. According to writer and educator Kendra Cherry, Gardner has suggested that "individuals who are high in this type of intelligence are more in tune with nature and are often interested in nurturing, exploring the environment, and learning about other species." Everyone, however, learns through nature to some degree, and Gardner's theory makes a compelling case that schools should use time in nature as a teaching tool and environment. Some children will be best served in this way, and all will have their learning enhanced.

In other words, playing outside at recess is not enough. We need to guide students to have direct, intimate experiences with nature and other animals — those deeply emotional "aha" moments that Joanne Vining and Melinda Merrick call "environmental epiphanies." Schools must incorporate "nature time" into the curriculum as part of the main school day. This can take the form of gardens, nature walks, reading outside, and youngsters telling stories about their experiences with animals and the outdoors. Two good books that can help are *Companions in Wonder*, edited by Julie Dunlap and Stephen Kellert, and *A Kids' Guide to Protecting & Caring for Animals*, by Cathryn Berger Kaye.

Another compelling reason to teach children to value other animals, so they do not think nonhuman animals are "less" than us, is that recent research shows that our attitudes about animals influence how we feel about humans. Brock University psychologists Gordon Hodson and Kimberly Costello have discovered that when children think animals are of lesser value than humans, this feeling of human superiority can lead to racial prejudice and other forms of discrimination. This includes dehumanizing attitudes against individuals of what are called "outgroups," including immigrants. This dehumanization predicts prejudice in adults. Valuing animals, then, is a win-win situation for all.

Examples of what it means to rewild education are everywhere. For fifteen years, I have taught a course at the Boulder County Jail that offers an opportunity for the inmates to begin to rewild with nature. We have frequent discussions about human-animal relationships and what the men would like to do when they get out. I know these discussions have made a difference because some of my former students have told me so: The course changed how they view and value nature. Similarly, the Cedar Creek Corrections Center in Littlerock, Washington, has a composting program that also gets inmates to reconnect with nature and learn about science. As one former prisoner said, "I went into prison and came out a scientist." We can also help to rehabilitate and rewild troubled youths and at-risk kids in programs such as those offered by Big City Mountaineers.

Ultimately, for me, rewilding education has three main goals, whatever shape it takes:

- Gaining knowledge about our interrelationships with animals puts our own impacts in perspective. Learning about who animals are and the threats to their existence helps us see the consequences of human actions and society.
- Learning about nonhuman animals drives home how much we depend on them and the vital role they play in our daily lives. Coexisting with and caring for other

animals is a necessity. Knowing about how their well-being and ours are closely tied together will help guide our decisions so we foster respect and coexistence.

- As children learn how sentient, intelligent, and even moral many animals are, they will naturally want to treat them better and more compassionately, for the animals' own sake. This moral imperative to be compassionate to others is certainly central to human social development, and it can be expressed and experienced very directly with animals.

These are practical, positive benefits of rewilding education and learning more about nature and the human-animal bond, and they make some the most urgent reasons for developing humane education programs.

Conservation Psychology

When *Psychology Today* readers question what my animal columns have to do with human psychology, I have two answers. One is to remind people that humans are also animals. The hearts and minds of other animals relate to us, and understanding nonhuman animals improves our understanding of how the hearts and minds of humans evolved and developed. But also, and more to the point, what we know, think, feel, and believe about animals, and the way that we treat them, *tells us lots about human psychology*.

As I often say, it is rarely a lack of knowledge and concrete data that result in the harm we cause to animals and ecosystems. Rather, losses to biodiversity are typically due to the inability of humans to come to terms with the notion of biodiversity or to understand their place in the ecology of Earth. Inadequate protection of the animals in our care is typically the result of human indifference and arrogance, to the belief that humans are superior and that other animals don't understand or deserve better treatment. Therefore, it is critical to address the important psychological, social, and cultural issues that support our poor stewardship of animals and their habitats and to take down the psychological barriers that prevent people from facing and addressing these complex, frustrating, and urgent human-induced problems. It is going to take a wide-ranging social movement to get us out of the incredible messes for which we are responsible.

This is the focus of the new and emerging field of conservation psychology, which includes conservation social work. Conservation psychology is defined as "the scientific study of the reciprocal relationships between humans and the rest of nature, with a particular focus on how to encourage conservation of the natural world." This is the "human dimension," and it includes why and how humans do some of the unconscionable and destructive things we do, as well as the constructive and positive actions we take. Why, for example, do people override their feelings of biophilia? Why do humans break appropriate regulations and laws to harm nature? What leads some people to develop particularly deep and compassionate connections to nature, and how can we inspire this in everyone?

In many ways, conservation psychology is the scientific face of what I mean by rewilding. A summary of research in this field can be found in Susan Clayton and Gene Myers's outstanding book called *Conservation Psychology: Understanding and Promoting Human Care for Nature* and Susan Clayton's *The*

Oxford Handbook of Environmental and Conservation Psychology. Changing our attitudes and beliefs about other animals really is a social movement, and Nick Cooney's book *Change of Heart: What Psychology Can Teach Us about Spreading Social Change* also shows how our beliefs inform our actions. For those who want more, Thomas Ryan's *Animals and Social Work: A Moral Introduction* is also an excellent book. If rewilding is to develop into a wider social movement, the ideas and lessons of conservation psychology will guide it.

Of course, people choose to harm nature for a wide variety of reasons, some of which have to do with personal survival, not a lack of compassion. Plus, the usual collection of selfish motivations and differences can lead people to devalue animals and nature: financial gain, professional status, laziness, willful ignorance about the damage we cause, genuine disagreements about the extent of the threats facing us, and so on. At the local level, laws are often broken because there is a complete lack of awareness of existing regulations. When communities are impoverished, and survival leaves them no choice but to value their own welfare over their ecosystem's, then social justice goes hand in hand with ecological welfare.

We must discuss the loss of animal species directly and openly as part of our wider discussions of social justice, human welfare, and ecological damage. We must face how human attitudes about nature and animals are a major influence. People may disagree about the exact extent of human impacts on species declines and extinctions, but the facts show clearly that those impacts are real. And there are many more losses than those to which we pay attention. Melanie Challenger rightly points out in her book *On Extinction* that we are not only losing species.

Languages, cultures, and ways of life are disappearing as well. We have really made some huge messes.

So, what do we need to do? We need an inclusive and cohesive social movement that attracts the attention and hearts of academics and experts across all disciplines — biologists, ethologists, psychologists, sociologists, anthropologists, economists, political scientists, philosophers, historians, lawyers, artists, writers, and more. Everyone can make a difference in their own area and across spheres, and these spheres support one another. On the one hand, getting animal issues on political agendas is critical for reaching a wide public audience, but widespread public attention and concern is also one effective way to inspire politicians to act. (For more on animal and conservation politics, see recent books by David Meyer, David Johns, Denise Russell, and Sue Donaldson and Will Kymlicka.)

A developing field of study called conservation social work at the Institute for Human-Animal Connection at the University of Denver can also play a large role in helping to smooth and improve our interactions with other animals. Social workers typically do not focus directly on nonhuman animals in their work, but by expanding their work to focus on the well-being of all family members, human and nonhuman, they can play a large part in the rewilding social movement. In *Animals and Social Work*, Thomas Ryan sums this up nicely: "This dependency [of animals on their human companions] demands human responsibility for their welfare and well-being.... Social workers have a special responsibility to the weak and vulnerable of *all* species; once it is acknowledged that animals are part and parcel of the moral fabric, the sooner we shall come to see that we have duties and obligations that extend beyond the confines of our own species. Indeed social workers of a century hence may well come to view with incredulity the fact that their predecessors ever failed to extend respect to sentient creatures, or chose to remain morally indifferent to their plight."

The notion of rewilding clearly is a great meeting place, a unifying concept with a broad agenda that can really have a significant social impact if we allow it to. We need to move ahead with less hubris and more — much more — humility and recognize just how powerful we are and how our well-being is closely tied into the well-being of other species and our planet as a whole. Rewilding will have effects at different scales, but we really do hold the future of the planet in our hearts, heads, hands, and tools. We are the only species that can really change things for the better. We are that powerful, and in that sense, we are that exceptional.

Afterword

Rewild as You Go

The most common way people give up power is by thinking they don't have any.

— Alice Walker

We humans became human through our interactions with other animals. Knowing this fact gives us a new way to think about our responsibilities to help our hurt natural world today.... As we save other animals, we'll be saving ourselves.

— Barbara King

For rewilding to become a transformative social movement, people from all walks of life, and with often vastly different interests, will need to work with one another. This will only happen if rewilding itself is regarded as a flexible concept that can be adapted to fit a wide range of different contexts and needs, and if the movement itself is undertaken from a spirit of humility, grace, kindness, compassion, and empathy.

This is my vision of rewilding. To succeed, we must be both hopeful and pragmatic, idealistic and realistic, persistent and flexible, and kind to, and critical of, one another. We need to rewild our hearts and build landscapes of hope and heart. We must live with hope, not in fear. We need to recognize that we are also animals and therein lies hope for a much better future. The resilience of other animals and nature as a whole is being tested and strained to its limits, and at some point the rubber band will likely break. What a tragedy that will be for all beings.

As any reader of my *Psychology Today* essays knows, I feel very strongly about animal protection and conservation, and I am very grateful to have a public forum in which to express my perspective and hopefully convince others of the intimate connection between their lives and the lives of all nonhuman animals. As a result of my column, I also come face to face on almost a daily basis with the full range of other opinions people have on these issues. As I said earlier, I've found that most people feel very passionately about animals, and about how humans impact nature generally, and even those who flat-out disagree with me often raise valid points and offer perspectives that deserve consideration.

Clearly, we are a very complex species, and we live in an amazingly beautiful, magnificent, and complex world that constantly challenges us on many different levels. We are also a very talented and compassionate species; many people do not feel comfortable with what humans have and are doing to the planet and about what this says about who we are. Yet I also feel we often make things more complex than they need to be. On the one hand, I know that the messes we've made are not at all simple: We struggle to understand what's happening and why, much less with how to fix the damage. On the other hand, I think if everyone made some simple changes to their lives, the world would soon become a more compassionate place for all beings and landscapes. At root, it's about adopting the attitude that you can never be too kind or too generous or too compassionate, for these attitudes are our most powerful tools for connecting with others and with our world (for instance, see *The Power of Kindness* by Piero Ferrucci). Another vital attitude is to believe you can always do more. If we constantly challenge ourselves to do better within the context of our own lives, these good acts will radiate outward and transform our world.

The reality is that we are not all the same. People live in vastly different circumstances around the globe, and these circumstances often define what people can do. Many spend most of their time merely trying to survive, and their choices are limited. Others have more influence, affluence, and flexibility. People also have vastly different levels of caring. Some only become motivated to make changes to help nonhuman animals when their own lives are impacted, and in fact some may never be convinced to care about nonhuman animals or our environment. This is always hard for me to understand, but it's important to acknowledge.

Sure, I admit I would like the world to be vegan and for all animals, human and nonhuman, to have high-quality lives, to be able to live in peace and safety, absent of abuse and suffering. However, I am also a realist. This is not going to happen anytime soon, if ever. So the real question is: Can we humans — complex beings who we are — agree to add compassion to our daily activities in ways that rewild our hearts and then stick to this through thick and thin? We owe it to future generations to do so. We really need radical compassion and radical rewilding that become part of who we are and how we act in the world.

As I circle the globe and meet a vast array of people, I find myself more accepting of different lifestyles and tactics for making the world a better place for other animals. I've learned we need to talk *with* people, not *to* or *at* them, because we need all people to be involved. We need researchers and scientists to take an interdisciplinary approach. We also need as many everyday people and families as possible to embrace compassionate conservation as a daily mantra. As I have said, science alone is not going to make us more compassionate, and we don't need more data and information to convince us that action is needed. Moral decisions about who lives and who dies confront us right now, and the precautionary principle should lead us to address our wider environmental problems even without knowing every single cause.

In the best of all possible worlds, everyone would realize where we are at and do something about it. This may not yet be the best of all possible worlds, but it's really pretty darned good. One of the key ways to rewild is to concentrate on what works. If we value the central ethic of rewilding, then we just need to pursue it in our own lives as best we can, while doing whatever we can to convert the unconverted.

Discouragement, Stress, and Burnout

The state of our world, and the extent of the problems facing us, are sobering and can be overwhelming. For activists and anyone, it's vital to avoid burnout and the debilitating anxiety and grief that leads to inaction. In fact, rewilding ourselves is the perfect antidote. We need to have some balance. We need cheerful moments and to celebrate successes, and we need time in nature and with other animals just for ourselves. Bruce Gottlieb, a good friend and outstanding psychologist, has talked with me about "secondary trauma," which I believe can affect activists for animals and the environment. Also called "vicarious trauma," this refers to "the stress resulting from helping or wanting to help a traumatized or suffering person." Of course, the same stress can arise when working with nonhuman animals or even nature. In many ways, this trauma is what leads to burnout. No matter how hard people work, nor how many successes they have, there is always more suffering in the world.

I have often wondered why I haven't burned out despite many decades as an activist working for other animals. The reason, I have come to realize, is that I'm constantly rewilding. Every day I connect with nature and the animals around my home, and I hold to the unwavering belief that I'm doing some good in the world. I work really hard on a lot of "ugly stuff," but I nurture the resilience to keep at it by making sure my life is balanced: I've learned how to "get away from it all" for a while and return fully recharged. I believe in what I do, even if there isn't a gold star at the end of the day. Indeed, I may not live to see the fruits of my labor, but that's just fine.

I like to say I work hard, I play hard, and I rest hard. To get away from heady and depressing stuff, I often like to watch mindless movies or read mindless books and to sip good red wine and single-malt scotch. Sometimes I stir the scotch with Twizzlers. Those who know me well know that I'm addicted to watching tennis matches and cycling races over and over again. I don't need any more reality drama in these restful moments. I really turn off my brain, or as Bruce would put it, I walk away from my cortex and leave it behind for a while. Among my compassionate mother's last words to me was to be sure to play a lot, and I also laugh at myself, which is something my dear father told me long ago was essential for good mental health.

Animal suffering by human hands exists in all corners of the world, and we may never stop or end all of it. Yet I remain convinced that what others and I are doing is making positive, real progress. To maintain this attitude day to day, I realize that I often compartmentalize what is happening so that I can step away from brain, pain, and trauma as necessary. While what I do may not work for everyone, here is how I approach my work, finding balance and rewilding everyday, and how I keep the faith that we can indeed make the world a better place for all beings and their homes. For more discussion and advice, see *Trauma Stewardship* by Laura Van Dernoot Lipsky.

This is a modified version of the eight Ps of rewilding in chapter 3:

- Think positively. Don't let people or "bad" situations get you down. I'm not a 100-percent blind optimist, but negativity is a time and energy suck. Remind yourself of the good things that are happening and rekindle your faith from time to time. Take deep breaths and do something you enjoy. The bottom line is take care of yourself. This is the most important thing to remember; otherwise you won't have the energy to continue.
- Caring for animals, nature, or our planet is not radical, and it doesn't make you the "bad guy." You don't have to apologize for feeling compassion or working for a better world. In fact, fighting to end callous indifference and needless harm to animals and Earth is heroic work.
- Seek to find areas of common ground. It is imperative

that we work together as much as possible in the ways we can. To coexist with other species and retain the integrity of ecosystems, humanity must act as a unified collective. Often there aren't quick fixes or individual solutions.

- Be vocal and have the courage to speak out. We need to encourage everyone to act as concerned citizens and responsible stewards. We must lead by example.
- Be proactive. Look at your life, or at what is happening in your community, and actively change what can be changed. Look at what can be improved. Don't simply run around always "putting out the fires" that have started.
- Appeal to compassion with compassion. Be nice and kind to those with whom you disagree and move on if necessary. Sometimes it's better to let something go or agree to disagree. In other words, pick your battles carefully, and don't waste finite and valuable time and energy on people or situations you can't influence or that won't change. Some people love to fight and have no interest in finding solutions. Instead of engaging them, turn your attention to another way to help animals and Earth.

Stay Positive

Compassion begets compassion, and there is a synergistic relationship, not a trade-off, when we show compassion for animals and their homes. There are indeed many reasons for hope; for inspiration, see *Reason for Hope* by Jane Goodall and *This*

Is Hope by Will Anderson. Plus, according to Tali Sharot in *The Optimism Bias,* there is compelling evidence that we are natural-born optimists. And so, I'd like to end this book with a very positive vision and assessment of where we are. Many people around the world are working hard to reverse the harm for which we are responsible, and there are many reasons to hope that as a society we can harness our basic goodness and optimism and all work together as a united community.

In his latest book, *The Nature Principle*, Richard Louv paraphrases iconic civil rights leader Martin Luther King Jr. as follows: Any cultural movement will fail if it can't paint a picture of a world where people want to go to. This is sage advice for everyone working hard to make the world a more peaceful and compassionate place for all beings. I'm thrilled to be part of an interdisciplinary and international program called Obstacles and Catalysts of Peaceful Behavior. Among the goals of this program is to develop a "science of peace." This is not only important for the survival of other animals but also for the survival of humanity.

For starters, read Paul Chappell's book *Peaceful Revolution* and Peter Diamandis and Steven Kotler's book *Abundance: The Future Is Better than You Think*, both of which I have enjoyed and which made me think even more about keeping my own dreams and hopes alive. Erle Ellis also argues that, despite the problems we've caused, we may have more time than we think, since we can't know if the Earth really is approaching an irreversible tipping point. Dr. Ellis writes, "To deny the likelihood of an impending global tipping point is not to deny that we are transforming the biosphere profoundly and permanently in ways that are likely to disgrace us in the eyes of future generations. Much of our planet's ecology can and will be lost unless we focus much greater effort on conserving and restoring it.... The claim that the biosphere is approaching a global tipping point remains no more than a contested and untested hypothesis. As we strive towards more sustainable stewardship of our planet, we must think globally — but let us not lose track of problems on smaller scales. The fate of the entire biosphere depends on it."

I really like the phrase "leap and the net will appear," the idea put forth by American naturalist John Burroughs. We need to have faith that what we do will have positive effects. In many ways we intuitively know what we need to do, but we can be afraid to commit to something when we are unsure of what the results will be. Instead, we keep our hopes and dreams alive by taking the step into action, doing something, anything, based on what our heads and hearts are telling us. We need to step out of our comfort zones and think out of the box and work with others. We need a new mind-set of cooperating with others who also care about animals and Earth. Hard work pays off. A report issued in May 2012 by the Center for Biological Diversity showed that about 90 percent of endangered species are recovering on time. The report notes that we still have a long way to go to know how well many species are doing, but there are success stories across the United States. These include the Aleutian Canada goose, California least tern, American crocodile, black-footed ferret, whooping crane, gray wolves in the northern Rocky Mountains, and the shortnose sturgeon.

We humans are deeply entrenched in virtually every single web of nature, and if we think of us all as one, we will find ways to support many diverse beings in a single community. In *The Rational Optimist*, British scientist and journalist Matt

Ridley argues "that the world will pull out of its current crisis because of the way that markets in goods, services, and ideas allow human beings to exchange and specialise honestly for the betterment of all." He calls this view rational optimism. I hope he's right. Jeremy Rifkin, in *The Empathic Civilization*, also proposes that empathy will help us solve our various crises, and indeed, as I've mentioned, ample data show that both human and nonhuman animals are far more empathic and cooperative than previously thought.

As we move on, there will be give and take, and compromise and triage are not always easy or popular. It's better to dance than to fight because we can get a lot more done working together. *Rewilding as part of a global social movement can unify us once and for all.* Rewilding has many faces and can easily be a pivot point and a magnet — a paradigm change — not only for bringing diverse people together but also for reconnecting people with their own hearts. Alienating ourselves from other animals and dominating them and their homes is not what it means to be human. We must stop this insanity now. Ecocide is suicide. I started rewilding as a kid and haven't stopped. Maintaining hope and keeping our dreams alive requires a collective and selfless effort.

We live in a magnificent yet wounded world. Despite all of the rampant destruction and abuse, it remains a magnificent world filled with awe and wonder. If you're not in awe, you're not paying attention. So let's get on with it. Open your heart to nature and rewild as you go through your daily routines and rituals. The beginning is now. We can always do more as we rewild. Rewilding is a work in progress from which we must not get deflected. How lucky we are that we are able to partake in this process, gratefully and generously blurring borders between "them" and "us" and their homes and ours.

Let's make personal rewilding all the rage. We are all intimately interconnected, we are all one, and we all can and must work together as a united community to reconnect with nature and to rewild our hearts.

Acknowledgments

As he has for many years in the past, Jason Gardner has offered wonderful support, and I deeply appreciate his critical eyes and ears and good humor. Monique Muhlenkamp has been a wonderful publicist, and indeed, everyone else at New World Library — Ami Parkerson, Munro Magruder, Jonathan Wichmann, and Kristen Cashman — has been an absolute pleasure to work with. I smile and cringe when I listen to other writers tell me the horror stories they've endured working with other publishers. Not so for me. Jeff Campbell has once again been an awesome copyeditor and has an uncanny ability to read my mind and heart.

Discussions with Jane Goodall, Lori Gruen, Dale Jamieson, Colin Allen, Michael Tobias, Jill Robinson, Sarah Bexell, Camilla Fox, Dan Ramp, Mark Derr, Yan Chun Su, Andy Pruitt, Giulia Buttarelli, Carron Meaney, Bruce Gottlieb, Ingrid Newkirk, and

Hope Ferdowsian have always kept my brain alive and kicking, and Valerie Belt always keeps me up to date with the latest and the greatest news. My longtime cycling buddies, Bill (Billy) and Connor Simmons, always have been and continue to be great sounding boards, and since they're faster than me on the bike, they can always ride away if I get too involved. They rarely do, so thanks for taking it easy on me. Jessica Pierce, my coauthor on *Wild Justice: The Moral Lives of Animals* and a variety of essays, also remains a wonderful colleague and friend who has also listened to my ideas over the years. Jessica can drop me like a rock, especially on tricky descents on a mountain bike, but she's been wonderfully patient, so I hope she thinks I'm on to something. A special thanks to all of the people who sent me stories and continue to do so.

I also thank Rod Bennison and his wonderful family for putting me up and for putting up with me on my many trips to Australia. It was in the peaceful environs of their home in Keinbah (near Lovedale, New South Wales) in the wine-laden Hunter Valley where many of my ideas about rewilding came to mind and to fruition. And last but surely not least, Bob Slater helped me set up my office in my mountain home so I could have reliable Internet, much needed in this age of never-ending information coming in at rates too high for my primitive brain to track. If I missed thanking anyone, I deeply apologize, as I've met so many wonderful people around the world who have inspired me to keep on going...and going.

Endnotes

Introduction: What Is "Rewilding Our Hearts"?

Page 1, *"When human beings lose their connection"*: Chögyam Trungpa, *Shambhala: The Sacred Path of the Warrior* (Boston: Shambhala, 1988), 125.

Page 4, *"Our world is becoming global"*: Pat Shipman, *The Animal Connection: A New Perspective on What Makes Us Human* (New York: W. W. Norton & Co., 2011), 280.

Page 5, *"Many of the students who have crossed"*: Paul Ehrlich, *A World of Wounds: Ecologists and the Human Dilemma* (Oldendorf/Luhe, Germany: Ecology Institute, 1997), 15.

Page 5, *In the most basic sense, "rewilding" means:* See J. B. MacKinnon, *The Once and Future World: Nature as It Was, As It Is, As It Could Be* (Boston and New York: Houghton Mifflin Harcourt, 2013).

Page 8, *"Rewilding is all the rage in conservation circles"*: Editorial, *New Scientist* (March 1, 2014), 5.

Page 8, *two well-known conservation biologists, Michael Soulé and Reed Noss:* Michael Soulé and Reed Noss, "Rewilding and Biodiversity: Complementary Goals for Continental Conservation," *Wild Earth* 8 (1998), 19–28.

Page 9, *In her book* Rewilding the World, *conservationist Caroline Fraser:* Caroline Fraser, *Rewilding the World: Dispatches from the Conservation Revolution* (New York: Metropolitan Books, 2009), 9.

Page 9, *That is, according to Dave Foreman — the director:* See the Rewilding Institute, http://rewilding.org/rewildit. See also two books by Dave Foreman: *Rewilding North America: A Vision for Conservation in the 21st Century* (Washington, DC: Island Press, 2004) and *Man Swarm and the Killing of Wildlife* (Durango, CO: Raven's Eye Press, 2011).

Page 9, *One such ambitious, courageous, and forward-looking effort:* For more about the Yellowstone to Yukon Conservation Initiative, visit http://y2y.net.

Page 9, *ecological rewilding efforts center on the difficult question:* for a discussion of this question, see Lawrence J. Cookson, "A Definition for Wildness," *Ecopsychology* 3 (September 2011), 187–93.

Page 11, *Often called the "Paul Revere of Ecology":* For an overview of Barry Commoner's life and work, see Daniel Lewis, "Scientist, Candidate and Planet Earth's Lifeguard," *New York Times,* October 1, 2012, http://www.nytimes.com/2012/10/02/us/barry-commoner-dies-at-95.html.

Page 12, *"A human being is a part of the whole":* This quote is from a 1950 letter written by Albert Einstein to Robert Marcus; see http://en.wikiquote.org/wiki/Albert_Einstein.

Page 13, *We need to "be humble in the face of nature's awesomeness":* Eva Emerson, editorial, *Science News* (June 29, 2013), http://www.sciencenewsdigital.org/sciencenews/20130629?pg=4#pg4.

Page 13, *For example, Andrew Balmford and Richard Cowling stress:* Andrew Balmford and Richard M. Cowling, "Fusion or Failure: The Future of Conservation Biology," *Conservation Biology* 3 (June 2006), 692–95, DOI: 10.1111/j.1523-1739.2006.00434.x.

Page 13, *California State University psychologist P. Wesley Schultz notes:* P. Wesley Schultz, "Conservation Means Behavior," *Conservation Biology* 25 (2011), 1,080–83.

Page 15, *"Conservation is humankind's attempt":* Mark Hoelterhoff, "New York State of Mind," *Psychology Today* (April 27, 2012), http://www.psychologytoday.com/blog/second-nature/201204/new-york-state-mind.

Page 15, *it has been suggested that we rebuild rather than rewind:* "Rebuilding Not Rewinding Is the Future," editorial, *New Scientist*, no. 2917 (May 16, 2013), http://www.newscientist.com/article/mg21829173.000-rebuilding-not-rewinding-is-the-future-of-conservation.html#.U2f6jBygZjx.

Page 16, *Erle Ellis, who works in the Department of Geography:* Quotes by Erle Ellis from the following: Erle Ellis, "The Planet of No Return," *The Breakthrough* (winter 2012), http://thebreakthrough.org/index.php/journal/past-issues/issue-2/the-planet-of-no-return; and Erle Ellis, "Forget Mother Nature: This Is a World of Our Making," *New Scientist*, no. 2816 (June 14, 2011), 26–27, http://www.newscientist.com/article/mg21028165.700-forget-mother-nature-this-is-a-world-of-our-making.html.

Page 16, *Renowned environmentalist Bill McKibben says:* Bill McKibben, *Eaarth: Making a Life on a Tough New Planet* (New York: Henry Holt, 2010), 16.

Chapter 1: Global Problems, Personal Unwilding

Page 21, *"Bereft of contact with wildness, the human mind":* David Abram, cover endorsement for Peter H. Kahn Jr. and Patricia H. Hasbach, eds., *The Rediscovery of the Wild* (Cambridge, MA: MIT Press, 2013).

Page 21, *"We have conquered the biosphere and laid waste":* Edward O. Wilson, *The Social Conquest of Earth* (New York: Liveright Publishing Corporation, 2012), 13.

Page 22, *Animals aren't "ghosts in our machine":* This phrase comes from the title of the movie *The Ghosts in Our Machine*, http://www.theghostsinourmachine.com.

Page 22, *others claim that in our rapidly changing world concepts such as "natural":* "Rebuilding Not Rewinding Is the Future of Conservation," editorial, *New Scientist*, no. 2917 (May 16, 2013), http://www.new

scientist.com/article/mg21829173.000-rebuilding-not-rewinding-is-the-future-of-conservation.html#.U2f6jBygZjx.

Page 22, *one suggestion has been made to create a "world park" using:* See Society of Ethnobiology, Shawn Sigstedt, "World Park; an Exciting New Global Nature Conservation Strategy and Education Plan," http://ethnobiology.org/world-park-exciting-new-global-nature-conservation-strategy-and-education-plan.

Page 23, *"Environmental problems have contributed to numerous collapses":* Paul Ehrlich and Anne Ehrlich, "Can a Collapse of Global Civilization Be Avoided?" *Proceedings of the Royal Society B* 280, no. 1754 (March 7, 2013), http://rspb.royalsocietypublishing.org/content/280/1754/20122845.full.

Page 24, *Overpopulation is a key factor in species extinction:* For more about this, visit the website of the Center for Biological Diversity, "Population and Sustainability," http://www.biologicaldiversity.org/programs/population_and_sustainability/index.html.

Page 24, *I like how Warren Hern, a local Boulder physician, puts it:* Warren Hern, "Guest Opinion: Population 7 Billion," *Daily Camera*, November 6, 2011, http://www.dailycamera.com/ci_19266032._

Page 25, *the Center for Biological Diversity is now giving out condoms:* "100,000 Endangered Species Condoms Shipped to 50 States," Center for Biological Diversity, press release, September 23, 2011, http://www.biologicaldiversity.org/news/press_releases/2011/condoms-09-23-2011.html. For more on the relation of environmentalism and overpopulation, read Michael Tobias, "Pro-Planet Is Pro-Choice," *Forbes* (May 9, 2011), http://www.forbes.com/sites/michaeltobias/2011/05/09/pro-planet-is-pro-choice/.

Page 26, *"We're certainly a dominant species, but that's not the same":* Paul Larmer, "History's Lesson: Build Another Noah's Ark," interview with Michael Soulé, *High Country News* 24, no. 9 (May 13, 2002), http://www.hcn.org/issues/226/11219.

Page 26, *"Ecosystems are not only more complex than we think":* Quoted in Dave Foreman, "Wilderness: From Scenery to Nature," in J. Baird Callicott and Michael P. Nelson, eds., *The Great New Wilderness Debate* (Athens: University of Georgia Press, 1998), 580.

Page 26, *there are strong indications that we have seriously underestimated:* Michael Le Page, "Climate Downgrade: Arctic Warming," *New Scientist*, no. 2891 (November 14, 2012) 34–39, http://www.newscientist.com/article/mg21628911.500-climate-downgrade-arctic-warming.html.

Page 26, *In November 2012, scientists in my hometown of Boulder, Colorado:* Charlie Brennan, "Boulder Scientists Highlight Record Polar Melting," *Daily Camera*, November 29, 2012, http://www.dailycamera.com/science-environment/ci_22090311/study-polar-ice-sheets-antarctica-greenland-melting-3.

Page 28, *Indeed, individuals of innumerable species are struggling to adapt:* For more on how species are adapting to climate change, see the following: Aaron Kunz, "Which Mammals Will Adapt to Climate Change?" Boise State Public Radio, May 14, 2012, http://earthfix.opb.org/flora-and-fauna/article/which-mammals-will-adapt-to-climate-change; Bryan Walsh, "The New Age of Extinction," *Time* (April 2, 2009), http://content.time.com/time/specials/packages/article/0,28804,1888728_1888736_1888858,00.html; and Nancy Ross-Flanagan, "Animals on the Move," *Science News* (June 15, 2012), https://www.sciencenews.org/article/animals-move.

Page 28, *Specific examples of these impacts accumulate almost daily:* Details in this paragraph come from the following sources: Hannah S. Mumby, Alexandre Courtiol, Khyne U. Mar., and Virpi Lummaa, "Climatic Variation and Age-Specific Survival in Asian Elephants from Myanmar," *Ecology* 94, no. 5 (May 2013), http://dx.doi.org/10.1890/12-0834.1; Roberta Kwok, "Breaking a Sweat," *Conservation* (January 9, 2013), http://conservationmagazine.org/2013/01/breaking-a-sweat/; Roberta Kwok, "Girls Only," *Conservation* (April 1, 2013), http://conservationmagazine.org/2013/04/girls-only/; Kristina Chew, "Climate Change Has Created a Lobster Eat Lobster World," Care2.com (July 30, 2103), http://www.care2.com/causes/climate-change-has-created-a-lobster-eat-lobster-world.html; Zoe Cormier, "Starfish Sacrifice Arms to Beat the Heat," *New Scientist* (May 29, 2013), http://www.newscientist.com/article/dn23615-starfish-sacrifice-arms-to-beat-the-heat.html; Roberta Kwok, "Rest Stop

Closed," *Conservation* (April 30, 2013), http://conservationmagazine.org/2013/04/rest-stop-closed/; and Michael Marshall, "Up to Half of All Birds Threatened by Climate Change," *New Scientist*, no. 2922 (June 19, 2013), http://www.newscientist.com/article/mg21829224.000-up-to-half-of-all-birds-threatened-by-climate-change.html.

Page 29, *it's been discovered that ocean acidification can reverse the response of nerve cells:* Janet Raloff, "Acidification Alters Fish Behavior," *Science News* (February 25, 2012), http://www.sciencenewsdigital.org/sciencenews/20120225/.

Page 29, *it's been found that three-lined skinks, a type of lizard:* Alasdair Wilkins, "Global Warming Might Make Lizards Super-Intelligent," io9.com, January 12, 2012, http://io9.com/5875545/global-warming-might-make-lizards-super-intelligent.

Page 30, *According to the Center for Biological Diversity, we're experiencing:* Center for Biological Diversity, "The Extinction Crisis," http://www.biologicaldiversity.org/programs/biodiversity/elements_of_biodiversity/extinction_crisis/.

Page 30, *This loss influences the many habitats in which amphibians live:* Indian Institute of Science, Centre for Ecological Sciences, "Ecological Significance," http://ces.iisc.ernet.in/biodiversity/amphibians/ecological.htm.

Page 30, *In addition, the "biodiversity boom" in Madagascar:* The details in this paragraph come from the following sources: Roberta Kwok, "Madagascar's Biodiversity Boom Has Slowed," *Conservation* (July 9, 2013), http://conservationmagazine.org/2013/07/madagascars-biodiversity-boom-has-slowed/; Christopher Collins and Roland Kays, "Causes of Mortality in North American Populations of Large and Medium-sized Mammals," *Animal Conservation* 14 (October 2011), 1–10; Michael Marshall, "Chimps Hunt Monkey Prey Close to Local Extinction," *New Scientist* (May 12, 2011), http://www.newscientist.com/article/dn20469-chimps-hunt-monkey-prey-close-to-local-extinction.html; and Jessica Junker, et al., "Recent Declines in Suitable Environmental Conditions for African Great Apes," *Diversity and Distribution* (November 2012), 1–15, DOI: 10.1111/ddi.12005.

Page 31, *a new species of bird, called the Cambodian tailorbird:* Roberta Kwok,

"New Bird Species Found Near Cambodian City," *Conservation* (July 1, 2013), http://conservationmagazine.org/2013/07/new-bird-species-found-in-cambodian-city/.

Page 31, *In 2012, among the new animal species discovered were a butterfly:* Beth Buczynski, "10 Amazing New Animal Species Discovered in 2012," Care2.com, December 12, 2012, http://www.care2.com/causes/10-amazing-new-animal-species-discovered-in-2012.html.

Page 31, *Chimpanzees in Uganda's Kibale National Park work together to catch prey:* J. S. Lwanga, et al., "Primate Population Dynamics Over 32.9 years at Ngogo, Kibale National Park, Uganda," *American Journal of Primatology* 73 (May 2011), 1–15, DOI: 10.1002/ajp.20965. See also Michael Marshall, "Chimps Hunt Monkey Prey Close to Local Extinction," *New Scientist* (May 12, 2011), http://www.newscientist.com/article/dn20469-chimps-hunt-monkey-prey-close-to-local-extinction.html.

Page 32, *At a talk I heard by Cornell University's Christopher Clark in November 2012:* Christopher Clark and Brandon Southall, "Turn Down the Volume in the Ocean," *CNN*, January 20, 2012, http://www.cnn.com/2012/01/19/opinion/clark-southall-marine/index.html.

Page 32, *in a* New Scientist *article entitled "Unnatural Selection":* Michael Le Page, "Unnatural Selection: How Humans Are Driving Evolution," *New Scientist*, no. 2810 (April 30, 2011), 32–37, http://www.newscientist.com/special/unnatural_selection.

Page 35, *Glenn Albrecht, professor of sustainability at Murdoch University:* Marc Bekoff, *The Emotional Lives of Animals* (Novato, CA: New World Library, 2007), 163.

Page 36, *conservation biologist Michael Soulé calls the climate change deniers:* Michael Soulé, email message to author, September 24, 2012.

Page 37, *one recent US survey found that "fewer than 10 percent":* Mark van Vugt and Vlada Griskevicius, "Let's Use Evolution to Turn Us Green," *New Scientist*, no. 2882 (September 21, 2012), http://www.newscientist.com/article/mg21528820.200-lets-use-evolution-to-turn-us-green.html.

Page 39, *such as the massacre of forty-nine captive wild animals in Ohio in October 2011:* Marc Bekoff, "Bloodbath in Ohio: Numerous Exotic

Animals Killed After Being Freed," *Psychology Today* (October 19, 2011), http://www.psychologytoday.com/blog/animal-emotions/201110/bloodbath-in-ohio-numerous-exotic-animals-killed-after-being-freed.

Page 39, *it has been estimated that, each year, 37 to 120 billion farmed fish:* "Fish Count Estimates," Fishcount.org.uk, http://fishcount.org.uk/fish-count-estimates.

Page 41, *But all sorts of evidence continues to emerge that kindness:* Details in this paragraph come from the following sources: Marco Schmidt and Jessica Sommerville, "Fairness Expectations and Altruistic Sharing in 15-Month-Old Human Infants," *PLoS ONE* (October 7, 2011), DOI: 10.1371/journal.pone.0023223; Jessica Pierce and Marc Bekoff, "Moral in Tooth and Claw," *The Chronicle of Higher Education* (October 18, 2009), http://chronicle.com/article/Moral-in-ToothClaw/48800/; and Wilson, *Social Conquest of Earth.*

Page 42, *This is true for nonhuman animals as well:* Cara Santa Maria, "Animal Morality Research Suggests We All Have Complex Emotions," *HuffPost Science,* September 10, 2012, http://www.huffingtonpost.com/2012/09/10/animal-morality-research_n_1859579.html.

Page 42, *"Gus, as most dogs are, was fiercely territorial about his yard":* Diane Snider, email message to author, July 1, 2012.

Page 42, *And there's mounting evidence that being compassionate is good:* Elizabeth Svoboda, "How Neuroscience Can Make You Kinder," *New Scientist,* no. 2939 (October 23, 2013), http://www.newscientist.com/article/mg22029390.300-how-neuroscience-can-make-you-kinder.html.

Page 42, *Conservation psychologist Susan Clayton at the College of Wooster reports:* Susan Clayton, "The Point of Public Lands," *Psychology Today* (March 11, 2012), http://www.psychologytoday.com/blog/being-green/201203/the-point-public-lands.

Chapter 2: Compassion and Coexistence Mean It's Not All about Us

Page 45, *"If we don't always start from nature, we certainly":* Henry Miller, *Big Sur and the Oranges of Hieronymus Bosch* (New York: New Directions, 1957), 93.

Page 45, *"In the traditions of the many different Native peoples":* Joseph Bruchac, "Native Americans' Relationships with Animals: All Our Relations," in Marc Bekoff, ed., *Encyclopedia of Animal Rights and Animal Welfare* (Westport, CT: Greenwood Press, 1998), 254–55.

Page 47, *"our human evolutionary birthright to be the dominant animal":* Barbara King, "Anthropological Perspectives on Ignoring Nature" in Marc Bekoff, ed., *Ignoring Nature No More: The Case for Compassionate Conservation* (Chicago: University of Chicago Press, 2013), 205.

Page 51, *"We have created the human framework for the problem":* Philip Tedeschi, email message to author, May 23, 2013. For more on the Institute for Human-Animal Connection, see www.humananimal connection.org.

Page 52, *in early 2014, the Copenhagen Zoo killed a young healthy male giraffe:* Marc Bekoff, "Healthy Young Zoo Giraffe to Be Killed: 'Zoothanasia' Redux," *Psychology Today* (February 8, 2014), http://www.psychologytoday.com/blog/animal-emotions/201402/healthy-young-zoo-giraffe-be-killed-zoothanasia-redux; and *The Guardian*, "Danish Zoo that Killed Marius the Giraffe Puts Down Four Lions," March 25, 2014, http://www.theguardian.com/world/2014/mar/25/danish-copenhagen-zoo-kills-four-lions-marius-giraffe.

Page 52, *We must be committed to looking for the most humane solutions:* for more on this, see also: Sara Dubois and H. W. Harshaw, "Exploring 'Humane' Dimensions of Wildlife," *Human Dimensions of Wildlife* 18 (2013), 1–19.

Page 53, *Research now proposes that morality and compassion are "rooted":* Mark Rowlands, "The Kindness of Beasts," *Aeon Magazine* (October 24, 2012), http://aeon.co/magazine/being-human/mark-rowlands-animal-morality/.

Page 53, *"The blind eye we turn to the suffering of animals":* Richard Foster, email message to author, January 21, 2012; see also the *Daily Kumquat*, www.dailykumquat.com.

Page 53, *the best examples of shared caring between different species:* Marc Bekoff, "Odd Couples: Compassion Doesn't Know Species Lines," *Psychology Today* (October 28, 2012), http://www.psychology

today.com/blog/animal-emotions/201210/odd-couples-compassion-doesnt-know-species-lines.

Page 54, *"Your remarks on elephants reminded me of a short article":* Marius Donker, email message to author, September 25, 2012.

Page 55, *Almut Beringer, from the University of Prince Edward Island in Canada, describes:* Almut Beringer, "The 'Spiritual Handshake': Toward a Metaphysical Sustainability Metrics," *Canadian Journal of Environmental Education* 12 (2007), 143–59, http://cjee.lakeheadu.ca/index.php/cjee/article/viewFile/622/519.

Page 57, *"More science and more technology are not going to get":* Lynn White, "The Historic Roots of Our Ecologic Crisis," *Science* 155 (1967), 1203–12.

Page 57, *"There is a difference... between the call of the outdoors":* Kathleen Dean Moore, endorsement for Peter H. Kahn Jr. and Patricia H. Hasbach, eds., *The Rediscovery of the Wild* (Cambridge, MA: MIT Press, 2013).

Page 58, *"The majority of our most deeply held beliefs are immune":* Michael Shermer, *The Believing Brain* (New York: Times Books, Henry Holt and Co., 2011), 4.

Page 58, *Joe Zammit-Lucia says poignantly, "Conservation":* Joe Zammit-Lucia, "Conservation Is Not about Nature," *International Union for Conservation of Nature,* September 7, 2011, http://www.iucn.org/involved/opinion/?8195/Conservation-is-not-about-nature.

Page 59, *The perspective offered by David Haskell, an ecologist:* James Gorman, "Finding Zen in a Patch of Nature," *New York Times,* October 22, 2012, http://www.nytimes.com/2012/10/23/science/david-haskell-finds-biology-zen-in-a-patch-of-nature.html.

Page 61, *studies are increasingly showing that connecting with nature is:* The studies in this paragraph are described in the following sources: Brian Clark Howard, "Connecting with Nature Boosts Creativity and Health," *National Geographic Daily News,* June 28, 2013, http://news.nationalgeographic.com/news/2013/06/130628-richard-louv-nature-deficit-disorder-health-environment/; David Suzuki Foundation, "Connect with Nature to Reduce Stress, *CBC,* http://www.cbc.ca/liverightnow/tips-and-articles/30x30/connect

-with-nature-to-reduce-stress.html; and Martin Dallimer, et al., "Biodiversity and the Feel-Good Factor: Understanding Associations between Self-Reported Human Well-being and Species Richness," *BioScience* 62 (January 2012), 47–55.

Page 62, *"[The] results are striking in that they demonstrate that compassion":* David DeSteno, "The Power of Compassion as a Moral Force," *Psychology Today* (March 11, 2011), http://www.psychologytoday.com/blog/out-character/201103/the-power-compassion-moral-force.

Chapter 3: Making It Real: Hard Choices and Bottom Lines

Page 65, *"No one wants to see polar bears, or pandas":* Susan Clayton, ed., *The Oxford Handbook of Environmental and Conservation Psychology* (New York: Oxford University Press, 2012).

Page 73, *we know that being positive and hopeful are important for getting people to care:* Michael Slezak, "How to Convince Climate Sceptics to Be Pro-Environment," *New Scientist*, no. 2870 (June 20, 2012), http://www.newscientist.com/article/mg21428702.300-how-to-convince-climate-sceptics-to-be-proenvironment.html; Adam Corner, "Climate Science: Why the World Won't Listen," *New Scientist*, no. 2936 (September 26, 2013), http://www.newscientist.com/article/mg21929360.200-climate-science-why-the-world-wont-listen.html.

Page 74, *it has been suggested that positive media — showing an elephant with her offspring:* Andrew Adam Newman, "Avoiding Violent Images for an Anti-Poaching Campaign," *New York Times*, February 19, 2013, http://www.nytimes.com/2013/02/20/business/media/world-wildlife-fund-anti-poaching-campaign-avoids-violent-images.html.

Page 74, *We also need to be persistent and stick like glue to our beliefs:* For more on a positive, pragmatic approach to environmental activism, see Joe Zammit-Lucia, "Environmentalism Refreshed: P2," *Stanford Social Innovation Review* (January 9, 2013), http://www.ssireview.org/blog/entry/environmentalism_refreshed_part_2; and Elin Kelsey, "Ecologists Should Learn to Look on the Bright Side," *New Scientist*, no. 2846 (January 10, 2012), http://www.newscientist.com/article/mg21328460.200-ecologists-should-learn-to-look-on-the-bright-side.html.

Page 75, *"Welcome to the weird world of green economics"*: Fred Pearce, "Costing the Earth: The Value of Pricing the Planet," *New Scientist*, no. 2888 (October 31, 2012), http://www.newscientist.com/article/mg21628882.000-costing-the-earth-the-value-of-pricing-the-planet.html.

Page 77, *A study by researchers from the University of Groningen:* Roberta Kwok, "How to Be Good," *Conservation* (March 8, 2013), http://conservationmagazine.org/2013/03/how-to-be-good.

Page 78, *government programs (such as in Sweden) that financially reimburse ranchers:* Jason G. Goldman, "Predator Conservation Relies on Understanding Human Psychology," *Conservation* (May 2, 2014), http://conservationmagazine.org/2014/05/predator-conservation-relies-on-understanding-human-psychology.

Page 80, *Elizabeth Bennett points out that we really have not been very successful:* Elizabeth L. Bennett, "Another Inconvenient Truth: The Failure of Enforcement Systems to Save Charismatic Species," *Oryx* 45 (2012), 476–79.

Page 80, *Even young people, one report said, "place a higher priority":* Roberta Kwok, "All Things Bright," *Conservation* (January 23, 2013), http://conservationmagazine.org/2013/01/all-things-bright.

Page 80, *a single bee can contain twenty-five different agrochemicals:* Harry Eyres, "The Birds and the Bees," *Financial Times*, April 22, 2011, http://www.ft.com/cms/s/2/6534f050-6ba0-11e0-93f8-00144feab49a.html.

Page 81, *In a 2011 survey, which questioned 583 conservation:* Murray Rudd, "Scientists' Opinions on the Global Status and Management of Biological Diversity," *Conservation Biology* 25, no. 6 (December 2011), 1165–75, DOI: 10.1111/j.1523-1739.2011.01772.x.

Page 81, *a good approach is modeled by researchers working in Asia:* Wildlife Conservation Society, "WCS Releases List of Asian Species at the Conservation Crossroads," press release, September 5, 2012, http://www.wcs.org/press/press-releases/jeju-conservation-crossroads.aspx.

Page 81, *Successful reintroductions, also called repatriations, include putting wolves*: Details from the following sources: Smithsonian National

Zoological Park, "Golden Lion Tamarin Conservation Program," http://nationalzoo.si.edu/scbi/endangeredspecies/gltprogram/inwild/reintroduction.cfm; and Wildlife Extra, "Success and Failure of Releasing Animals Back into the Wild," http://www.wildlifeextra.com/go/news/animal-release739.html.

Page 84, *ecotourism can have positive economic benefits that help preserve wilderness:* Details in this paragraph are from the following sources: Ralf Buckley, "Endangered Animals Caught in the Tourist Trap," *New Scientist*, no. 2886 (October 15, 2012), http://www.newscientist.com/article/mg21628860.200-endangered-animals-caught-in-the-tourist-trap.html; Craig B. Stanford, *Planet Without Apes* (Cambridge, MA: Harvard University Press, 2012); and Jeffrey Gettleman, "To Save Wildlife, and Tourism, Kenyans Take Up Arms," *New York Times*, December 29, 2012, http://www.nytimes.com/2012/12/30/world/africa/to-save-wildlife-and-tourism-kenyans-take-up-arms.html.

Page 85, *in a very interesting and novel study, the Centre for Integral Economics:* Z. Parker and R. Gorter, "Crossroads: Economics, Policy, and the Future of Grizzly Bears in British Columbia," Centre for Integral Economics, June 2003, http://www.raincoast.org/files/publications/reports/Crossroads.pdf.

Page 85, *legalizing the hunting of predators as a conservation strategy — on the theory:* Jason G. Goldman, "Predator Conservation Relies on Understanding Human Psychology," *Conservation* (May 2, 2014), http://conservationmagazine.org/2014/05/predator-conservation-relies-on-understanding-human-psychology.

Page 87, *"As a resident of Montana I've often been confronted with the controversy":* Jessica B, "Montana Resident," November 2, 2011, comment on Marc Bekoff, "Rampant Wolf Killing Makes Some People Happy," *Psychology Today* (November 1, 2011), http://www.psychologytoday.com/blog/animal-emotions/201111/rampant-wolf-killing-makes-some-people-happy/comments.

Page 89, *One inspirational approach has been developed by a Kenyan teen:* Teo Kermeliotis, "Boy Scares Off Lions with Flashy Invention," *CNN*, February 26, 2013, http://www.cnn.com/2013/02/26/tech/richard-turere-lion-lights. See also, Richard Turere, "My Invention That

Made Peace with Lions," TED2013, February 2013, http://www.ted.com/talks/richard_turere_a_peace_treaty_with_the_lions.html.

Page 90, *As conservation biologists Adrian Treves and Jeremy Bruskotter rightly argue:* Jason G. Goldman, "Predator Conservation Relies on Understanding Human Psychology," *Conservation* (May 2, 2014), http://conservationmagazine.org/2014/05/predator-conservation-relies-on-understanding-human-psychology.

Page 90, *As I've mentioned, human well-being in general is enhanced:* For details in this paragraph and more information on biophilic cities, see the following sources: Art Markman, "Do Parks Make People Happier?" *Psychology Today* (June 26, 2013), http://www.psychologytoday.com/blog/ulterior-motives/201306/do-parks-make-people-happier; Ingrid L. Stefanovic and Stephen B. Scharper, eds., *The Natural City: Re-envisioning the Built Environment* (Toronto: University of Toronto Press, 2012); Biophilic Cities, http://biophiliccities.org; and Rewilding Europe, www.rewildingeurope.com.

Page 91, *I have had the pleasure of working with Timothy Beatley:* Timothy Beatley and Marc Bekoff, "City Planning and Animals: Expanding Our Urban Compassion Footprint," in Claudia Basta and Stefano Moroni, eds., *Ethics, Design, and Planning of the Built Environment* (New York: Springer, 2013).

Page 91, *I am proud to live in Boulder, Colorado, a truly biophilic city:* Victoria Derr and Krista Lance, "Biophilic Boulder: Children's Environments that Foster Connections to Nature," *Children, Youth and Environment* 22, no. 2 (2012), 112–43.

Page 92, *Toronto has been identified as one of the world's most deadly cities for birds:* Details on birds and cities come from the following sources: Ian Austen, "Casualties of Toronto's Urban Skies," *New York Times*, October 27, 2012, http://www.nytimes.com/2012/10/28/world/americas/casualties-of-torontos-urban-skies.html; Susan Milius, "Collision Course," *Science News* (September 5, 2013), https://www.sciencenews.org/article/collision-course; Roberta Kwok, "False Moon," *Conservation* (May 2, 2013), http://conservationmagazine.org/2013/05/false-moon; Susan Milius, "Noise May

Disrupt a Bat's Dinner," *Science News* (August 7, 2013), https://www.sciencenews.org/article/noise-may-disrupt-bat's-dinner; and the Fatal Light Awareness Program, www.flap.org.

Page 92, *"There is no substitute for reimmersing people in the world":* David Johns, "How Might Ecologists Make the World Safe for Biodiversity Without Getting Fired?" *Bulletin of the British Ecological Society*, no. 4 (2012), 51–53.

Page 92, *a 2012 news story titled "Wolves and Mountain Lions":* Jill Reilly, "Wolves and Mountain Lions 'Poised to Invade Densely Populated Cities in the United States,'" *Daily Mail*, October 5, 2012, http://www.dailymail.co.uk/news/article-2213440/Wolves-mountain-lions-poised-invade-densely-populated-cities-United-States.html.

Page 94, *As Timothy Egan notes in a discussion of the dangers:* Timothy Egan, "Sometimes the Bear Gets You," *New York Times*, September 29, 2011, http://opinionator.blogs.nytimes.com/2011/09/29/sometimes-the-bear-gets-you.

Page 94, *in April 2011, Bolivia announced it would grant all nature:* John Vidal, Bolivia Enshrines Natural World's Rights with Equal Status for Mother Earth," *The Guardian*, April 10, 2011, http://www.theguardian.com/environment/2011/apr/10/bolivia-enshrines-natural-worlds-rights.

Page 94, *In 2012, Colombia banned the illegal trade of night monkeys:* Angela Maldondo, "Colombia Bans the Night Monkey Trade!" *International Primate Protection League News*, December 2012, http://www.ippl.org/gibbon/wp-content/uploads/2010/09/IPPL_News-Dec2012.pdf.

Page 95, *"Howard has worked here since 2000 and was the very first bear worker":* Jill Robinson, email message to author, December 13, 2011.

Page 98, *Leilani Münter, a race car driver with a degree in biology:* For more about Leilani Münter, visit http://leilanimunter.com.

Page 98, *Other uncommon messengers include Linda Tucker:* For more information on the people in this paragraph, see the following sources: for Linda Tucker, see Michael Charles Tobias, "One Woman's Remarkable Quest to Save Africa's White Lion," *Forbes* (June 23, 2013), http://www.forbes.com/sites/michaeltobias/2013/06/23

/one-womans-remarkable-quest-to-save-africas-white-lion; for Francisco Mayoral, see Homero Aridjis, "Savior of the Whales," *New York Times*, October 31, 2013, http://www.nytimes.com/2013/11/01/opinion/savior-of-the-whales.html; for Howard Lyman, visit www.madcowboy.com.

Chapter 4: Rewilding the Media: Our Mirror Up to Nature

Page 101, *Since June 2009, I have written regular online essays for* Psychology Today: See my "Animal Emotions" column at *Psychology Today*, http://www.psychologytoday.com/blog/animal-emotions.

Page 102, *one study has shown that accurate information about climate change:* Susan Clayton and Gene Myers, *Conservation Psychology: Understanding and Promoting Human Care for Nature* (Oxford, UK: Wiley-Blackwell, 2009), 23, 152.

Page 105, *As Georgia State University's Carrie Packwood Freeman points out in her essay:* Carrie Packwood Freeman, "This Little Piggy Went to Press: The American News Media's Construction of Animals in Agriculture," *Communication Review* 12 (2009), 78–103, http://scholarworks.gsu.edu/communication_facpub/9.

Page 106, *There have been only two fatal wolf attacks on humans: NBC News*, "Fatal Wolf Attack Unnerves Alaska Village," March 17, 2010, http://www.nbcnews.com/id/35913715/ns/us_news-life/t/fatal-wolf-attack-unnerves-alaska-village/#.U2krjRylUw1.

Page 108, *A recent example concerns Tai, the female elephant star:* For more, see Marc Bekoff, "Elephant Abuse in Film: 'Water for Elephant' Star, Tai, Shocked with Electric Prod," *Psychology Today* (May 12, 2011), http://www.psychologytoday.com/blog/animal-emotions/201105/elephant-abuse-in-film-water-elephant-star-tai-shocked-electric-prod.

Page 108, *Another example is the 2012 film* The Hobbit: An Unexpected Journey: *Los Angeles Times*, "Quick Takes: 'Hobbit' Animal Deaths," November 20, 2012, http://articles.latimes.com/2012/nov/20/entertainment/la-et-quick-20121120.

Page 110, *"Not only am I concerned about the welfare and exploitation":*

Thomas Mangelsen, "Point of View: Game Farm Photography," Naturescapes.net, January 19, 2009, http://www.naturescapes.net/articles/conservation/point-of-view-game-farm-photography.

Page 111, *There are excellent organizations teaching and encouraging responsible:* For more on the Center for Environmental Filmmaking, see http://www.american.edu/soc/cef/About.cfm.

Page 111, *These include the 2013 documentaries* Blackfish *and* The Ghosts in Our Machine: For more on *Blackfish*, see www.blackfishmovie.com, and for *The Ghosts in Our Machine*, www.theghostsinourmachine.com. For more on director Liz Marshall (www.lizmars.com) and photographer Jo-Anne McArthur (www.joannemcarthur.com), visit their websites.

Page 113, *in October 2012, a video of chimpanzees tormenting and tossing around: CBS News*, "Zoo Chimps Torment Wayward Racoon," October 17, 2012, http://www.cbsnews.com/videos/zoo-chimps-torment-wayward-raccoon.

Page 114, *A similar thing happened in a 2009 National Public Radio (NPR) report:* Robert Krulwich, "Ants That Count!" *NPR*, November 25, 2009, http://www.npr.org/blogs/krulwich/2011/06/01/120587095/ants-that-count.

Page 114, *"Without a doubt, wildlife biologists, who are professionally trained,":* Todd Wilkinson, "As a Federal Agent, Carter Niemeyer Killed Wolves for a Living," *Wildlife Art Nature Journal* (November 2, 2011), http://www.wildlifeartjournal.com/articles/wildlife-art-journal-premium-content/autumn-2011/336/as-a-federal-agent-carter-niemeyer-killed-wolves-for-a-living.html.

Page 115, *A few weeks before, a* Psychology Today *cover story entitled:* Rebecca Webber, "Are You with the Right Mate?" *Psychology Today* (January 1, 2012), http://www.psychologytoday.com/articles/201112/are-you-the-right-mate.

Page 116, *Even the Lincoln Park Zoo in Chicago, Illinois, asked for a company: Associated Press*, "Zoo: Chimp Ads Desensitize Viewers," January 31, 2012, http://espn.go.com/chicago/nfl/story/_/id/7524698/lincoln-park-zoo-wants-super-bowl-chimp-ads-pulled-air.

Page 116, *"the public is less likely to think that chimpanzees are endangered":*

Stephen R. Ross, Vivian M. Vreeman, and Elizabeth V. Lonsdorf, "Specific Image Characteristics Influence Attitudes about Chimpanzee Conservation and Use as Pets," *PLoS ONE* (July 13, 2011), DOI:10.1371/journal.pone.0022050.

Page 116, *Another 2011 study by Kara Schroepfer and her colleagues:* Kara K. Schroepfer, Alexandra G. Rosati, Tanya Chartrand, and Brian Hare, "Use of 'Entertainment' Chimpanzees in Commercials Distorts Public Perception Regarding Their Conservation Status," *PLoS ONE* (October 12, 2011), DOI:10.1371/journal.pone.0026048.

Page 117, *Yet about five years later, it was reported that clownfish populations:* Richard Alleyne, "Demand for Real *Finding Nemo* Clownfish Putting Stocks at Risk," *The Telegraph*, June 26, 2008, http://www.telegraph.co.uk/earth/earthnews/3345594/Demand-for-real-Finding-Nemo-clownfish-putting-stocks-at-risk.html.

Chapter 5: Rewilding the Future: Wild Play and Humane Education

Page 119, *"If you don't know how to fix it, please stop breaking it":* YouTube, "Severn Suzuki Speaking at UN Earth Summit 1992," posted May 29, 2007, http://www.youtube.com/watch?v=uZsDliXzyAY.

Page 120, *As Jane Goodall puts it, we are stealing the future of our children:* John Aguilar, "Famed Primatologist Jane Goodall Drops in on Boulder's Whittier Elementary," *Daily Camera*, May 3, 2013, http://www.timescall.com/ci_23168404/famed-primatologist-jane-goodall-drops-boulders-whittier-elementary.

Page 120, *in March 2013, Valerie Belt, an animal activist and a grade-school teacher:* Valerie Belt, email message to author, March 16, 2013.

Page 121, *such as with Jane Goodall's global Roots & Shoots program:* Visit www.rootsandshoots.org.

Page 122, *In 2010, it was reported that kids in the United States spend:* For the studies mentioned in this paragraph, see the following sources: Tamar Lewin, "If Your Kids Are Awake, They're Probably Online," *New York Times*, January 20, 2010, http://www.nytimes.com/2010/01/20/education/20wired.html; and Richard Black, "Nature

Deficit Disorder 'Damaging Britain's Children,'" *BBC News*, March 29, 2012, http://www.bbc.com/news/science-environment-17495032.

Page 123, *A recent survey of children's books provided another indication:* Nanci Hellmich, "Study: New Children's Books Lack Reference to Nature, Animals," *USA Today*, February 28, 2012, http://usatoday30.usatoday.com/news/health/wellness/story/2012-02-27/Study-New-childrens-books-lack-reference-to-nature-animals/53275082/1.

Page 124, *"The binary opposition of nature and technology, outdoor and indoor":* Wendy Russell, email message to author, February 29, 2012.

Page 125, *"I am also concerned by the construction of terms such":* Joe Zammit-Lucia, email message to author, February 28, 2012.

Page 126, *"Nature experience isn't a panacea, but it does help children":* Hellmich, "Study: New Children's Books Lack Reference to Nature, Animals."

Page 126, *"Better a broken bone than a broken spirit":* For Play Wales, see www.playwales.org.uk/eng.

Page 126, *Mark Hoelterhoff, a psychologist at the University of Cumbria, has written:* Mark Hoelterhoff, "Parenting Confessional and the Importance of Green Play," *Psychology Today* (March 11, 2012), http://www.psychologytoday.com/blog/second-nature/201203/parenting-confessional-and-the-importance-green-play.

Page 127, *Psychologist Susan Linn notes, "Time in green space is essential":* Marc Bekoff, "Nature-Deficit Disorder Redux: Kids Need to Get Off Their Butts," *Psychology Today* (February 28, 2012), http://www.psychologytoday.com/blog/animal-emotions/201202/nature-deficit-disorder-redux-kids-need-get-their-butts.

Page 127, *Further, a study published in 2012 showed that "spending time in nature":* Roberta Kwok, "Wild Ideas," *Conservation* (December 13, 2012), http://conservationmagazine.org/2012/12/wild-ideas.

Page 128, *"to reemphasise that the growing body of scientific evidence confirming":* Bob Hughes, *Evolutionary Playwork* (London and New York: Routledge, 2001), 43–44.

Page 129, *Bio-outcomes include "an increase in brain size and organisation":* Hughes, *Evolutionary Playwork*, 324.

Page 129, *"If the activity is bounded by adult rules, if it is stiff":* Hughes, *Evolutionary Playwork*, 207, 325.

Page 129, *In earlier research, my colleagues Marek Spinka, Ruth Newberry, and I:* Marek Spinka, Ruth Newberry, and Marc Bekoff, "Mammalian Play: Training for the Unexpected," *Quarterly Review of Biology* 76 (2001), 141–68.

Page 129, *"So my wild playing child, as a representation of everything human":* Hughes, *Evolutionary Playwork*, 385.

Page 130, *As my colleague A. G. Rud says, we may need to "unschool":* A. G. Rud, *Albert Schweitzer's Legacy for Education: Reverence for Life* (New York: Palgrave Macmillan, 2011).

Page 130, *Today, much of this falls under the umbrella of what's called humane education:* For more information on humane education, see the following sources: A. G. Rud and Jim Garrison, eds., *Teaching with Reverence: Reviving an Ancient Virtue for Today's Schools* (New York: Palgrave Macmillan, 2012); Sarah Bexell, Olga S. Jarrett, and Xu Ping, "The Effects of a Summer Camp Program in China on Children's Knowledge, Attitudes, and Behaviors Toward Animals: A Model for Conservation Education," *Visitor Studies* 16 (2013), 59–81; Peter Verbeek, "Peace Ethology," *Behaviour* 145 (2008), 1,497–1,524; and Paul Waldau, *Animal Studies: An Introduction* (New York: Oxford University Press, 2013).

Page 131, *"individuals who are high in this type of intelligence":* Kendra Cherry, "Gardner's Theory of Multiple Intelligences," About.com, http://psychology.about.com/od/educationalpsychology/ss/multiple-intell_9.htm. For another overview, see "The Naturalistic Intelligence" by Bruce Campbell, Johns Hopkins School of Education, http://education.jhu.edu/PD/newhorizons/strategies/topics/mi/campbell.htm.

Page 131, *that Joanne Vining and Melinda Merrick call "environmental epiphanies":* Joanne Vining and Melinda S. Merrick, "Environmental Epiphanies: Theoretical Foundations and Practical Applications," in Clayton, *Oxford Handbook of Environmental and Conservation Psychology*, 485–508.

Page 131, *Brock University psychologists Gordon Hodson and Kimberly Costello:* Gordon Hodson and Kimberly Costello, "The Link Between

Devaluing Animals and Discrimination," *New Scientist*, no. 2895 (December 21, 2012), http://www.newscientist.com/article/mg21628950.400-the-link-between-devaluing-animals-and-discrimination.html.

Page 132, *Similarly, the Cedar Creek Corrections Center in Littlerock:* Devin Powell, "Prisons an Unlikely Laboratory," *Science News* (April 4, 2013), https://www.sciencenews.org/article/prisons-unlikely-laboratory.

Page 132, *in programs such as those offered by Big City Mountaineers:* See Big City Mountaineers, www.bigcitymountaineers.org.

Page 134, *It is going to take a wide-ranging social movement to get us out of the incredible messes:* Harold A. Mooney, Anantha Duraiappah, and Anne Larigauderie, "Evolution of Natural and Social Science Interactions in Global Change Research Programs," *Proceedings of the National Academy of Sciences* 110 (February 26, 2013), DOI: 10.1073/pnas.1107484110.

Page 136, *"This dependency [of animals on their human companions] demands human":* Thomas Ryan, *Animals and Social Work: A Moral Introduction* (New York: Palgrave Macmillan, 2011), 164–65.

Afterword: Rewild as You Go

Page 139, *"The most common way people give up power is by thinking":* From William Martin, *The Best Liberal Quotes Ever* (New York: Sourcebooks, 2004), 173.

Page 139, *"We humans became human through our interactions":* Barbara King, "An Anthropologist's View of the Anthropocene: Co-evolution Out of Balance?" in Bekoff, *Ignoring Nature No More*.

Page 143, *Bruce Gottlieb, a good friend and outstanding psychologist, has talked:* For more on "secondary trauma," see http://secondarytrauma.org.

Page 146, *I'm thrilled to be part of an interdisciplinary and international program:* "Obstacles and Catalysts of Peaceful Behavior" is run by the Lorentz Center, http://www.lorentzcenter.nl/lc/web/2013/527/info.php3?wsid=527&venue=Oort.

Page 146, *"To deny the likelihood of an impending global tipping point":* Erle

C. Ellis, "Time to Forget Global Tipping Points," *New Scientist*, no. 2907 (March 11, 2013), http://www.newscientist.com/article/mg21729070.200-time-to-forget-global-tipping-points.html.

Page 147, *the idea put forth by American naturalist John Burroughs:* Jeff Arnold, "Leap and the Net Will Appear," Zen Leadership blog, May 10, 2009, http://www.blog.zenleadership.net/2009/05/leap-and-net-will-appear.html.

Page 147, *A report issued in May 2012 by the Center for Biological Diversity:* Center for Biological Diversity, "Study: 90 Percent of Endangered Species Recovering on Time," press release, May 17, 2012, http://www.biologicaldiversity.org/news/press_releases/2012/esa-success-05-17-2012.html.

Page 148, *"that the world will pull out of its current crisis because of the way":* Matt Ridley, *The Rational Optimist: How Prosperity Evolves* (New York: HarperCollins, 2010), 10.

Bibliography

Anderson, Will. *This Is Hope: Green Vegans and the New Human Ecology*. Winchester, UK: Earth Books, 2013.

Barash, David. *Buddhist Biology: Ancient Eastern Wisdom Meets Modern Western Science*. New York: Oxford University Press, 2013.

Basta, Claudia, and Stefano Moroni, eds. *Ethics, Design, and Planning of the Built Environment*. New York: Springer, 2013.

Beatley, Timothy. *Biophilic Cities: Integrating Nature into Urban Design and Planning*. Washington, DC: Island Press, 2010.

Bekoff, Marc. *The Animal Manifesto: Six Reasons for Expanding Our Compassion Footprint*. Novato, CA: New World Library, 2010.

———. *Animals at Play: Rules of the Game*. Philadelphia: Temple University Press, 2008.

———. *The Emotional Lives of Animals*. Novato, CA: New World Library, 2007.

———, ed. *Ignoring Nature No More: The Case for Compassionate Conservation*. Chicago: University of Chicago Press, 2013.

———. *Minding Animals: Awareness, Emotions, and Heart*. New York: Oxford University Press, 2002.

———. *Why Dogs Hump and Bees Get Depressed: The Fascinating Science of Animal Intelligence, Emotions, Friendship, and Conservation*. Novato, CA: New World Library, 2013.

Bekoff, Marc, and Sarah M. Bexell. "Ignoring Nature: Why We Do It, the Dire Consequences, and the Need for a Paradigm Shift to Save Animals and Habitats and to Redeem Ourselves." *Human Ecology Review* 17 (2010): 70–74.

Bekoff, Marc, and Jessica Pierce. *Wild Justice: The Moral Lives of Animals.* Chicago: University of Chicago Press, 2009.

Berry, Thomas. *The Great Work: Our Way into the Future*. New York: Broadway Books. 1999.

Berry, Robert J. *God's Book of Works*. London: Continuum, 2003.

Billick, Ian, and Mary V. Price, eds. *The Ecology of Place*. Chicago: University of Chicago Press, 2011.

Burt, Jonathan. *Animals in Film*. London: Reaktion Books, 2002.

Cafaro, Philip and Eileen Crist, eds. *Life on the Brink: Environmentalists Confront Overpopulation*. Athens, GA: University of Georgia Press, 2012.

Candolin, Ulrika, and Bob Wong, eds. *Behavioural Responses to a Changing World: Mechanisms and Consequences*. Oxford, UK: Oxford University Press, 2012.

Carson, Rachel. *Silent Spring*. New York: Fawcett Crest, 1964.

Challenger, Melanie. *On Extinction: How We Became Estranged from Nature*. London: Granta Books, 2011.

Chappell, Paul. *Peaceful Revolution: How We Can Create the Future Needed for Humanity's Survival*. Westport, CT: Easton Studio Press, 2012.

Clayton, Susan, ed. *The Oxford Handbook of Environmental and Conservation Psychology*. New York: Oxford University Press, 2012.

Clayton, Susan, and Gene Myers. *Conservation Psychology: Understanding and Promoting Human Care for Nature*. Oxford, UK: Wiley-Blackwell, 2009.

Cooney, Nick. *Change of Heart: What Psychology Can Teach Us about Spreading Social Change*. New York: Lantern Books, 2011.

Corbey, Raymond, and Annette Lanjouw, eds. *The Politics of Species: Reshaping Our Relationships with Other Animals*. New York: Cambridge University Press, 2013.

Corning, Peter. *The Fair Society*. Chicago: Chicago University Press, 2011.

De Waal, Frans. *The Age of Empathy: Nature's Lessons for a Kinder Society*. New York: Three Rivers Press, 2009.

Diamandis, Peter H., and Steven Kotler. *Abundance: The Future Is Better Than You Think*. New York: Free Press, 2012.

Donaldson, Sue, and Will Kymlicka. *Zoopolis: A Political Theory of Animal Rights*. New York: Oxford University Press, 2011.

Dunlap, Julie, and Stephen R. Kellert, eds. *Companions in Wonder: Children and Adults Exploring Nature Together*. Cambridge, MA: MIT Press, 2012.

Ehrlich, Paul. *A World of Wounds: Ecologists and the Human Dilemma*. Oldendorf/Luhe, Germany: Ecology Institute, 1997.

Ehrlich, Paul, and Robert E. Ornstein. *Humanity on a Tightrope: Thoughts on Empathy, Family, and Big Changes for a Viable Future*. New York: Rowman & Littlefield Publishers, 2010.

Evernden, Neil. *The Natural Alien: Humankind and Environment*. Toronto: University of Toronto Press, 1999.

Ferrucci, Piero. *The Power of Kindness: The Unexpected Benefits of Leading a Compassionate Life*. New York: Jeremy P. Tarcher, 2006.

Foreman, Dave. *Man Swarm and the Killing of Wildlife*. Durango, CO: Raven's Eye Press, 2011.

———. *Rewilding North America: A Vision for Conservation in the 21st Century*. Washington, DC: Island Press, 2004.

———. *Take Back Conservation*. Durango, CO: Raven's Eye Press, 2012.

Fraser, Caroline. *Rewilding the World: Dispatches from the Conservation Revolution*. New York: Metropolitan Books, 2009.

Goodall, Jane. *Reason for Hope: A Spiritual Journey*. New York: Grand Central Publishing, 2000.

Goodall, Jane, and Marc Bekoff. *The Ten Trusts: What We Must Do to Care for the Animals We Love*. San Francisco: HarperCollins, 2002.

Gruen, Lori. *Entangled Empathy*. New York: Lantern Books, 2014.

Gullone, Eleonora. *Animal Cruelty, Antisocial Behavior, and Aggression: More than a Link*. Houndmills, UK: Palgrave Macmillan, 2012.

Harré, Niki. *Psychology for a Better World: Strategies to Inspire Sustainability*. New Zealand: University of Auckland, 2011. Available at psych.auckland.ac.nz/psychologyforabetterworld.

Hughes, Bob. *Evolutionary Playwork*. London and New York: Routledge, 2001.

Johns, David. *A New Conservation Politics: Power, Organization Building, and Effectiveness*. Hoboken, NJ: Wiley-Blackwell, 2009.

Kahn, Peter. H., Jr. *The Human Relationship with Nature: Development and Culture*. Cambridge, MA: MIT Press, 1999.

Kahn, Peter H., Jr., and Patricia H. Hasbach, eds. *The Rediscovery of the Wild*. Cambridge, MA: MIT Press, 2013.

Kaye, Cathryn Berger. *A Kids' Guide to Protecting & Caring for Animals*. Minneapolis: Free Spirit Publishing, 2012.

Kellert, Stephen. *Birthright: People and Nature in the Modern World*. New Haven, CT: Yale University Press, 2012.

Keltner, Dacher. *Born to Be Good: The Science of a Meaningful Life*. New York: W. W. Norton & Co., 2009.

Lipsky, Laura van Dernoot. *Trauma Stewardship: An Everyday Guide to Caring for Self While Caring for Others*. San Francisco: Berrett-Koehler Publishers, 2009.

Louv, Richard. *Last Child in the Woods: Saving Our Children from Nature-Deficit Disorder*. New York: Algonquin Books, 2005.

———. *The Nature Principle: Human Restoration and the End of Nature-Deficit Disorder*. New York: Algonquin Books, 2011.

Lutts, Ralph H. *The Nature Fakers: Wildlife, Science & Sentiment*. Golden, CO: Fulcrum Publishing Company, 1990.

MacKinnon, J. B. *The Once and Future World: Nature as It Was, As It Is, As It Could Be*. Boston and New York: Houghton Mifflin Harcourt, 2013.

McKibben, Bill. *Eaarth: Making a Life on a Tough New Planet*. New York: Times Books, Henry Holt, 2010

Melson, Gail. *Why the Wild Things Are: Animals in the Lives of Children*. Cambridge, MA: Harvard University Press, 2001.

Meyer, David. *The Politics of Protest: Social Movements in America*. New York: Oxford University Press, 2007.

Mitman, Gregg. *Reel Nature: America's Romance with Wildlife on Film*. Cambridge, MA: Harvard University Press, 1999.

Morin, Edgar, and Anne Brigitte Kern. *Homeland Earth*. Cresskill, NJ: Hampton Press, 1999.

Myers, O. E., Jr. *The Significance of Children and Animals: Social Development and Our Connections to Other Species*. 2nd rev. ed. West Lafayette, IN: Purdue University Press, 1998/2007.

Nagy, K., and Phillip David Johnson, eds. *Trash Animals: How We Live with Nature's Filthy, Feral, Invasive, and Unwanted Species*. Minneapolis: University of Minnesota Press, 2013.

Niemeyer, Carter. *Wolfer: A Memoir*. Boise, ID: BottleFly Press, 2010.

Oppenlander, Richard A. *Comfortably Unaware: Global Depletion and Food Responsibility*. Minneapolis: Langdon Street Press, 2011.

Pacelle, Wayne. *The Bond: Our Kinship with Animals, Our Call to Defend Them.* New York: William Morrow, 2011.

Palmer, Chris. *Shooting in the Wild: An Insider's Account of Making Movies in the Animal Kingdom*. San Francisco: Sierra Club Books, 2010.

Peterson, Anna. *Being Animal: Beasts and Boundaries in Nature Ethics.* New York: Columbia University Press, 2013.

Ridley, Matt. *The Rational Optimist: How Prosperity Evolves*. New York: HarperCollins Publishers, 2010.

Rifkin, Jeremy. *The Empathic Civilization: The Race to Global Consciousness in a World in Crisis*. New York: Jeremy P. Tarcher, 2010.

Rud, A. G. *Albert Schweitzer's Legacy for Education: Reverence for Life*. New York: Palgrave Macmillan, 2011.

Rud, A. G., and Jim Garrison, eds. *Teaching with Reverence: Reviving an Ancient Virtue for Today's Schools.* New York: Palgrave Macmillan, 2012.

Russell, Denise. *Who Rules the Waves? Piracy, Overfishing, and Minding the Oceans*. New York: Pluto Press, 2010.

Ryan, Thomas. *Animals and Social Work: A Moral Introduction*. New York: Palgrave Macmillan, 2011.

Saylan, Charles, and Daniel T. Blumstein. *The Failure of Environmental Education (And How We Can Fix It)*. Berkeley: University of California Press, 2011.

Sharot, Tali. *The Optimism Bias: A Tour of the Irrationally Positive Brain*. New York: Pantheon Books, 2011.

Shermer, Michael. *The Believing Brain: From Ghosts and Gods to Politics and Conspiracies — How We Construct Beliefs and Reinforce Them as Truths.* New York: Times Books, Henry Holt, 2011.

Shipman, Pat. *The Animal Connection: A New Perspective on What Makes Us Human*. New York: W. W. Norton & Co., 2011.

Specter, Michael. *Denialism: How Irrational Thinking Hinders Scientific Progress, Harms the Planet, and Threatens Our Lives.* New York: Penguin Press, 2009.

Sponsel, Leslie E. *Spiritual Ecology: A Quiet Revolution*. New York: Praeger, 2012.

Stanford, Craig B. *Planet Without Apes.* Cambridge, MA: Harvard University Press, 2012.

Stefanovic, Ingrid L., and Stephen B. Scharper, eds. *The Natural City: Re-envisioning the Built Environment.* Toronto: University of Toronto Press, 2012.

Steiner, Gary. *Anthropocentrism and Its Discontents: The Moral Status of Animals in the History of Western Philosophy.* Pittsburgh, PA: University of Pittsburgh Press, 2005.

Sterba, Jim. *Nature Wars: The Incredible Story of How Wildlife Comebacks Turned Backyards into Battlegrounds.* New York: Crown Publishers, 2012.

Stolzenburg, William. *Where the Wild Things Were: Life, Death, and Ecological Wreckage in a Land of Vanishing Predators.* New York: Bloomsbury, 2009.

Taylor, Paul. *Respect for Nature: A Theory of Environmental Ethics.* Princeton, NJ: Princeton University Press, 1986.

Van Dernoot Lipsky, Laura, with Connie Burk. *Trauma Stewardship: An Everyday Guide to Caring for Self while Caring for Others.* San Francisco: Berrett-Koehler Publishers, 2011.

Waldau, Paul. *Animal Studies: An Introduction.* New York: Oxford University Press, 2013.

Washington, Haydn. *Human Dependence on Nature: How to Help Solve the Environmental Crisis.* New York: Earthscan, 2013.

Williams, Terry Tempest. *Finding Beauty in a Broken World.* New York: Pantheon, 2008.

Wilson, Edward O. *The Social Conquest of Earth.* New York: Liveright Publishing Corporation, 2012.

Wuerthner, George, Eileen Crist, and Tom Butler, eds. *Keeping the Wild: Against the Domestication of the Earth.* San Francisco: Foundation for Deep Ecology, 2014.

Index

Abram, David, 21, 79
Abundance (Diamandis and Kotler), 146
acidification, 29
activism, 94–99
ADHD, 127
Agassiz, Louis, 123
"aha" moments, 131
Albrecht, Glenn, 35
Allen, Colin, 27
Allen, Marjorie, Baroness of Hurtwood, 126
Almada (Portugal), 98
Amazon River, 83
American Humane Association, 108
amphibians, 30, 49, 80
Amsterdam (Netherlands), 91
Anderson, Will, 145–46
angling, 85
animal activists, 94–99
animal cruelty
  in bear-bile industry, 95
  in entertainment industry, 39, 111
  during filmmaking, 108–9
  inconsistent reactions to, 39
  news media portrayals of, 102, 113–14
  during scientific research, 99, 114
  species loss as, 29
Animal Cruelty, Antisocial Behavior, and Aggression (Gullone), 102
animal emotions, 1–2, 48, 133
Animal Manifesto, The (Bekoff), 3, 122
animals. *See* nonhuman animals

Animals and Social Work (Ryan), 135, 136–37
Animals Asia, 6, 95
animal sentience
- animals as "ghosts" and, 112
- children's education about, 133
- hunting and, 86
- meat diet and, 104–5
- minimization of harm and, 72
- scientific acceptance of, 1–2, 48–49

Animals in Film (Burt), 106
animal studies, 130
Antarctica, 83
Anthropocene epoch, 16–17, 47
anthropocentrism, 47
Anthropocentrism and Its Discontents (Steiner), 47
anthropology, 7
anthropomorphism, 105–6
anthrozoology, 130
ants, 114
aquariums, 39, 111
Arctic National Wildlife Refuge, 43
ATVs, 84
Australia, 31
Avatar (film; 2009), 109

balance, right to, 94
Balmford, Andrew, 13
bats, and urban "sky glow," 92
BBC, 123
bear-bile industry, 95–96, 97
bears
- black, 65, 66, 67–68, 81–82
- grizzly, 85, 86
- moon, 6, 95–96, 97
- polar, 28, 79–80

Beatley, Timothy, 91
Behavioural Responses to a Changing World (Candolin and Wong), 10
Belt, Valerie, 120–21
Bennett, Elizabeth, 80
Beringer, Almut, 55
Berman, Marc, 56
Berry, Robert, 22
Berry, Thomas, 54, 77
Big City Mountaineers, 132
Billick, Ian, 17–18
biodiversity
- "boom," in Madagascar, 30
- human inability to come to terms with, 133–34
- human survival dependent on, 28, 29–30
- rebuilding attempts, 10, 15

biodiversity loss, 3
- causes of, 31–33, 133–34
- certainty of, 81
- compassionate conservation and, 79–83
- human impact of, 29–30, 135–36
- rate of, 30
- Three Rs approach to, 81–82
- *See also* species extinction

biology, 7
bio-outcomes, 129
biophilia, 34, 40, 120, 134
biophilic cities, 90–94
biosphere, tipping point of, 146–47

birds
  death of, from impact with buildings, 92
  migratory, 28
birth control, 25–26
Birthright (Kellert), 61
Blackfish (documentary film; 2013), 111
Blumstein, Daniel, 58
bobcats, 67
Bolivia, 94
borders
  redefining, 12–13
  use of term, 13
Boulder (CO), 14, 65–70, 91
Boulder County Jail, 132
brain, 56–57
Brazil, 31, 81
breeding programs, 52
British Columbia (Canada), 85
Brock University, 131
Bruchac, Joseph, 45
Bruskotter, Jeremy, 90
Buckley, Ralf, 84
Buddhism, 46
burnout, dealing with, 142–45
Burroughs, John, 103, 147
Burt, Jonathan, 106
busy-ness, 34
Butler, Tom, 23
butterflies, 31

Cafaro, Philip, 24
California State University, 13–14
Candolin, Ulrika, 10
carnivores, 9
Carson, Rachel, 29
Cedar Creek Corrections Center (Littlerock, WA), 132
Center for Biological Diversity, 25, 147
Center for Environmental Filmmaking (Washington, DC), 111
Centre for Integral Economics, 85
Challenger, Melanie, 135–36
Change of Heart (Cooney), 135
Chappell, Paul, 146
Chengdu (China), 6, 95–96
Cherry, Kendra, 131
Chicago (IL), 91
children
  as environmental activists, 119–20
  humane education of, 130–33
  play and, 126–30
  rewilding and, 122
  unwilding and, 120–21, 122–26
children's books, unwilding of, 123–24
chimpanzees, 31, 109, 113, 115–17
China, 26, 95–96, 97
circuses, 39, 98
cities, biophilic, 90–94
Clark, Christopher, 32
Clayton, Susan, 42–43, 65, 134–35
clean air/water, rights to, 94
climate change, 3
  animals affected by, 28–29
  birth control and, 26
  deniers of, 27–28, 36–37
  "Noah's ark" attitude toward, 29
  underestimated rate of, 26–27

clownfish, 117
coexistence, 45, 51, 61, 72, 132–33, 144–45
cognitive abilities, 62, 128
cognitive ethology, 19
College of Wooster, 42–43
Colombia, 94–95
Colorado, 85
Comfortably Unaware (Oppenlander), 36
commonality, seeking areas of, 144–45
Commoner, Barry, 11, 99
common good, the, 82–83
communications technology, 123, 127
companion animals
  endangered species as, 116
  human relationships with, 53
  language used with, 104
Companions in Wonder (ed. Dunlap and Kellert), 131
compassion, 4
  animal play behavior and, 128
  in animals, 42, 53–54
  apologizing for, 5, 144
  compassion resulting from, 43, 62, 145
  good feelings resulting from, 60–63
  human redecoration of nature without, 71–72
  moral imperative for, 133
  rewilding based in, 4–8, 38, 45–46, 139–42
  unwilding and loss of, 35
compassionate conservation, 7, 19, 78–83
computers, 127
condoms, 25
Congo, Democratic Republic of, 31
conservation
  behavior modification needed for, 14
  compassionate, 7, 19, 78–83
  media portrayals of endangered species and, 115–17
  rewilding and, 8–12
Conservation (journal), 127
conservation biology, 19, 37
conservation education, 130
conservation politics, 136
conservation psychology, 121, 134–37
Conservation Psychology (Clayton and Myers), 134
conservation social work, 134, 136–37
Cooney, Nick, 135
cooperation, 128
Copenhagen Zoo, 52
coral reefs, 76–77
Corbey, Raymond, 15
corridors, 9, 10–11, 12, 92
Costa Rica, 95
Costello, Kimberly, 131–32
cougars, 65, 67, 92–93
Cowling, Richard, 13
cranes, whooping, 147
creativity, 62, 127
Crist, Eileen, 23, 24

crocodiles, American, 147
cultural stereotypes, 106
cultures, extinctions of, 136

Daily Kumquat, 53
Dalai Lama, 62
damselfish, Australian, 29
Death in the Afternoon (Hemingway), 41
deep ethology, 7, 55
deer, 115
dehumanization, 132
denial, 33, 35–37
Denialism (Specter), 36
DeSteno, David, 62
Diamandis, Peter, 146
discouragement, dealing with, 142–45
Disney, Walt, 105–6
dogs, 107–8
domestication, 107
Donaldson, Sue, 136
Donker, Marius, 53–54
doomsday thinking, 37–38
dualisms, false, 19
Dunlap, Julie, 131

Earth
 Anthropocene epoch, 16–17, 47
 "minding," 6
 tipping point of, 146–47
Earth Summit (Rio de Janeiro; 1992), 119–20
ecolodges, 83
ecology
 human inability to come to understand, 133–34
 of place, 18
 social justice and, 135–37
 spiritual, 55 (*see also* deep ethology)
Ecology of Place (Billick and Price), 17–18
economics
 appeals to, 38
 green, 75–78
 of hunting, 86
 moral choices based on, 85–86
ecosystems
 compassion and, 46
 economic value of, 75–78
 human damage inflicted on, 3, 32, 34, 71–72
 rebuilding attempts, 10, 15
 restoration of, and compassionate conservation, 78–83
ecotourism, 83–85
education
 conservation, 130
 environmental, 58
 "nature time" as part of, 131
 rewilding of, 120, 130–33
Egan, Timothy, 94
Egler, Frank, 26
Ehrlich, Anne, 23
Ehrlich, Paul, 5, 23, 45
Einstein, Albert, 12
elephant poaching, 74, 84

elephants
  abuse of, during filmmaking, 108
  activists working to prevent slaughter of, 98
  African, 53–54, 66, 84, 98
  Asian, and climate change, 28
  compassionate behavior of, 53–54
  ecotourism and, 84–85
  positive media portrayals of, 74
elk, 115
Ellis, Erle, 16–17, 146–47
Emotional Lives of Animals, The (Bekoff), 3
Empathic Civilization, The (Rifkin), 148
empathy, 7–8, 45, 47, 128, 148
  See also compassion
endangered species
  amphibians as, 30
  choices for recovery, 79–83
  ecotourism and, 84–85
  grey wolves as, 107
  media misrepresentations of, 107, 115–17
  recovery of, 147
Endangered Species Act, 107
England, 82
Entangled Empathy (Gruen), 47
environmental activists, 94–99
Environmental Children's Organization (ECO), 119–20
environmental education, 58
environmental epiphanies, 131
environmentally friendly products, 37
ethics
  of hunting, 86
  science and, 2
  with teeth, 82
ethology, 7, 19, 55
Evolutionary Playwork (Hughes), 128–30
evolutionary theory, 41–42
exercise, physical, 128
existence value, 43

Failure of Environmental Education, The (Saylan and Blumstein), 58
faith, 147
Fatal Light Awareness Program, 92
fear-mongering, 93, 106–7
ferrets, 82, 147
Ferrucci, Piero, 141
fidelity, rule of, 52
Finding Nemo (film; 2003), 117
Fine, Aubrey, 40
fish, 48
fishing, 85
flandry, 89
food, animals killed for, 39, 87–88, 105
Foreman, Dave, 9, 58
Foster, Richard, 53
four-wheeling, 84
foxes, 66, 67, 69–70, 72
Fox, Michael W., 99
Fraser, Caroline, 9
Freeman, Carrie Packwood, 105
Fromm, Erich, 34

game farms, nature photography at, 110–11
Gardner, Howard, 131
geese, Aleutian Canada, 147
genetic alteration, right to be free of, 94
Georgia State University, 105
Ghosts in Our Machine, The (documentary film; 2013), 111–12
global warming. *See* climate change
Global White Lion Protection Trust, 98
Goodall, Jane, 3, 27, 120, 121, 145
Gottlieb, Bruce, 143
Grand Canyon, 43, 83
Grand Teton National Park, 94
Great Work, The (Berry), 54
green economics, 75–78
greenhouse gases, 76
green play, 126–27
Grey, The (film; 2011), 106–7
Griffith University (Brisbane, Australia), 84
group selection, 41–42
Gruen, Lori, 47
guard dogs, 89
Gullone, Eleonora, 102
Gus (dog), 42

habitat destruction, 72
hamsters, 82
harm, minimizing, 70–72
Harré, Niki, 61
Haskell, David, 59–60
healing, 62
Hemingway, Ernest, 41
Hern, Warren, 24–25
"Historical Roots of Our Ecological Crisis, The" (White), 57
"hit men," 114
Hobbit, The (film; 2012), 108–9
Hodson, Gordon, 131–32
Hoelterhoff, Mark, 15, 127
Holocene epoch, 16
Homo denialus, 35–37
honeybees, 80
How Animals Talk (Long), 103
"hug a hunter" program, 85
Hughes, Bob, 128–30
human-animal relationship
  author's experience with, 19–20
  biophilic cities and, 91
  borders in, 12–13
  paradoxical nature of, 104
  study of, 130, 132
  unwilding and, 33–35
human beings
  ecological impact of, 3
  play behavior of, 123
  unique qualities of, 49–50
humane education, 121, 130–33
human exceptionalism, 34, 46–50, 132, 134
Humanity on a Tightrope (Ehrlich and Ornstein), 45
Human Relationship with Nature, The (Kahn), 122
hunting, 84, 85–87, 95

ice, melting of polar caps, 26–27
Idaho, 87
Idaho Springs (CO), 121–22
immigrants, dehumanization of, 132
insects, 30, 80
Institute for Human-Animal Connection (University of Denver), 51, 136
interconnectedness
  climate change and, 29
  ecology of place and, 18
  of ecosystems, 11
  education about, 132
  human exceptionalism vs., 48
  rewilding and importance of, 48, 55–56, 149
  spiritual background of, 46
  unwilding and inability to see, 35
interdependence, 46
International Centre for Tourism Research, 84
Internet, 104, 123
interspecies friendships, 53
intuition, 7
invasive species, 32

Jamaica, 31
Jaws (film; 1975), 107
Johns, David, 92, 136

Kahn, Peter, Jr., 122
Kaye, Cathryn Berger, 131
Keeping the Wild (ed. Wuerthner, Crist, and Butler), 23
Kellert, Stephen, 61, 131
Kenya, 84
Kibale National Park (Uganda), 31
Kids' Guide to Protecting & Caring for Animals, A (Kaye), 131
King, Barbara, 47, 139
King, Martin Luther, Jr., 146
kites, red, 82
Kotler, Steven, 146
Kymlicka, Will, 136

language
  extinctions of, 136
  use of, 102, 104–5
Lanjouw, Annette, 15
Last Child in the Woods (Louv), 122
Le Page, Michael, 32–33
Life on the Brink (ed. Cafaro and Crist), 24
lifestyles, extinctions of, 136
light pollution, 92
Lincoln Park Zoo (Chicago, IL), 116
Linn, Susan, 127
lions, 66
Lipsky, Laura Van Dernoot, 144
"Little Red Riding Hood," 106
London, Jack, 103
Long, William J., 103
Louisville (CO), 97
Louv, Richard, 122, 126, 146
Lutts, Ralph, 103
Lyberth, Angaangaq, 27
Lyman, Howard, 98

Madagascar, 30
mammals, large, 30–31
Mangelsen, Thomas, 110–11
Marius (giraffe), 52
Markman, Art, 90
Marshall, Liz, 111
Mayoral, Francisco, 98
McArthur, Jo-Anne, 112
McKibben, Bill, 16
media
    influence of, 102
    positivity in, 74
    sensationalism in, 102–4
media, animals as portrayed in
    animal actors, 108–9
    endangered species and conservation status, 115–17
    film/TV portrayals, 105–8
    nature documentaries/photography, 109–12
    news portrayals, 112–15
    objectification, 101
    sensationalism, 92–93, 103–4
Melson, Gail, 122
Merrick, Melinda, 131
Mexico, 98
Meyer, David, 136
migration, 28
Miller, Henry, 45, 56
"minding animals," 6–7
Minding Animals (Bekoff), 3
Missouri, 81–82
Mitman, Gregg, 106
monkeys, 31, 95
Montana, 87–88, 98
Monterey Bay (CA), 31
Moon Bear Rescue Centre (Chengdu, China), 6, 95–96
Moore, Kathleen Dean, 57
moose, 115
morality, 50–54, 85–86
Moss, Cynthia, 66
mountain lions. *See* cougars
multiple intelligences, 131
Münter, Leilani, 98
Murdoch University (Perth, Australia), 35
Myers, O. E. ("Gene"), 122, 134

National Public Radio (NPR), 114
National Snow and Ice Data Center (Boulder, CO), 26–27
Native Americans, 45
naturalistic intelligence, 131
nature
    in children's books, 124
    compassion and, 46
    enjoyment of, 83–87
    firsthand education about, 130
    human alienation from, 1–4, 33, 61 (*see also* unwilding)
    human control of, 81
    human footprint in, 21–22
    human redecoration of, 70–72
    population control through, 25
    positive effects of, 56–57, 60–63
    rights granted to, 94
    as "survival of the fittest," 110
    wildness vs. the outdoors, 57
    world park preserving, 22–23

nature corridors, 92
nature-deficit disorder, 122, 124–25
nature documentaries, 109–10, 111–12
Nature Fakers, The (Lutts), 103
nature photography, 110–12
Nature Principle, The (Louv), 146
Nature Wars (Sterba), 93
nature writing, 103
negativity, 73, 74, 144
Newberry, Ruth, 129
New Scientist, 8, 32–33, 75–77
New York Times, 60, 92
Niemeyer, Carter, 114–15
Noah (film; 2014), 109
noise pollution, 32, 92
nonhuman animals
 in children's books, 124
 climate change impact on, 28–29
 compassionate behavior of, 42, 53–54
 computer-generated, for filmmaking, 109
 dressed as humans, 115
 as "ghosts," 112
 illegal trade in, 95
 interspecies friendships, 53
 language used to refer to, 104–5
 "minding," 6–7
 objectification of, 101, 105
 play behavior of, 123, 127–28
 romanticizing, 5
 use of term, 3
 See also animal emotions; animal sentience; companion animals; human-animal relationship; media, animals as portrayed in; predatory animals; *specific animal species*
noninterference, rule of, 52
nonmaleficence, rule of, 52
North America, large-mammal loss in, 30–31
Northeastern University, 62
Noss, Reed, 8–9

objectification, 101–4
objectivity, 4, 6, 74
Obstacles and Catalysts of Peaceful Behavior, 146
ocean acidification, 29
ocean global commons, 32
one-child policies, 26
On Extinction (Challenger), 135–36
Oppenlander, Richard, 36
optimism, rational, 148
Optimism Bias, The (Sharot), 146
orcas, 111, 121
Oregon State University, 57
Ornstein, Robert, 45
outdoors, the
 fear of, 38
 wildness vs., 57
outfitters, 86–87
outgroups, dehumanization of, 132
overconsumption, 23–24, 37, 81

overpopulation, 3, 17, 21
birth control and, 25–26
ecological impact of, 24, 26, 81
as environmental challenge, 71
overconsumption and, 23–24
play and, 123
rate of increase, 24–25
Oxford Handbook of Environmental and Conservation Psychology, The (Clayton), 134–35

Pacelle, Wayne, 75
"pain points," 23
Palmer, Chris, 106
pandas, 79–80
passion, 72–73, 74
patience, 72–73, 75
peace, 7–8, 146
peacefulness, 72–73
Peaceful Revolution (Chappell), 146
Pearce, Fred, 75–77
persistence, 72–73, 74–75
philosophy, 7
Phnom Penh (Cambodia), 31
phones, 127
Pierce, Jessica, 82
pikas, American, 28
play
adult-supervised, 124–25, 128, 129
evolutionary playwork, 128–30
importance of, 123
"nature time" as part of, 131
rewilding of, 126–30
risk entailed in, 129
"Playing Into the Future—Surviving and Thriving" conference (Wales; 2011), 123
Play Wales (charity), 126
poaching, 74, 84
polar ice, melting of, 26–27
Politics of Species, The (ed. Corbey and Lanjouw), 15
pollution, 32, 34, 94
pop culture, 102
Portland State University, 92
positivity, 72–74, 144, 145–49
power, 72–73, 74
Power of Kindness, The (Ferrucci), 141
practicality, 72–73, 75
precautionary principle, 27–28, 59
predatory animals
controversies over, 87–90
hunting of, legalized, 85
killing of, as "self-defense," 114–15
recovery of, in urban environments, 93
sensationalistic media portrayals of, 92–93, 103–4
warding off, 89–90
Price, Mary, 18
proactivity, 72–73, 145
Psihoyos, Louie, 98
psychology, 7
Psychology for a Better World (Harré), 61
Psychology Today, 101, 104, 115, 133, 140

raccoons, 113
racial prejudice, 132
rainforests, 77
ranchers, and wolf reintroductions, 87–90
rational optimism, 148
Rational Optimist, The (Ridley), 147–48
Reason for Hope (Goodall), 145
Reel Nature (Mitman), 106
reintroductions, 81–82
religious beliefs, 38
Respect for Nature (Taylor), 52
restitutive justice, rule of, 52
rewilding
  author's experience with, 65–70, 143
  children and, 122
  compassion as foundation of, 4–8, 38, 45–46
  conservation psychology as scientific face of, 134
  as conservation strategy, 8–12
  defined, 5–6, 38, 40–41
  of economics, 75–78
  of education, 130–33
  Eight Ps of, 72–75, 144–45
  as flexible concept, 139
  good feelings resulting from, 15, 41
  human exceptionalism and, 46–50
  of language, 104–5
  main goal of, 70
  media and, 102
  moral choices based in, 50–54
  as natural, 40–43
  need for, 70
  as personal life strategy, 12–15, 65–70, 143
  of play, 126–30
  role models, 94–99
  science as insufficient for, 57–60
  selfish motives for, 30
  as spiritual movement, 54–57
  storytelling and, 107–8
  of urban landscapes, 90–94
  use of term, 124
  See also rewilding as social movement
"Rewilding and Biodiversity" (Soulé and Noss), 9
rewilding as social movement
  compassion as foundation of, 139–42
  conservation psychology and, 134–37
  dealing with burnout, 142–45
  positivity needed in, 145–49
  role models, 94–99
  unifying influence of, 148–49
Rewilding Europe Project, 91
Rewilding Institute (Albuquerque, NM), 9
rewilding projects, 9–12
Rewilding the World (Fraser), 9
Ridley, Matt, 147–48
rights, of nature, 94
Rise of the Planet of the Apes (film; 2011), 109

roadless areas, 9
Robinson, Jill, 95–96
rodeos, 39
romanticization, 5
Roosevelt, Theodore, 103
Roots & Shoots program, 121
Ross, Stephen, 116
Rud, A. G., 129
Russell, Denise, 136
Russell, Wendy, 124–25
Ryan, Thomas, 135, 136–37

Saylan, Charles, 58
Schroepfer, Kara, 116
Schultz, P. Wesley, 13–14
Schweitzer, Albert, 130
science/scientists
  animals killed for, 39, 114
  compassion vs., 141
  ethics and, 2
  human exceptionalism undermined by, 48–49
  as insufficient solution, 57–60, 63
  of peace, 146
  subjectivity and, 4–5
SeaWorld (Orlando, FL), 111, 121
sensationalism, 102–4, 106–7
Serengeti, the, 83
Seton, Ernest Thompson, 103
Sharot, Tali, 146
Sheldrake, Rupert, 103
Shermer, Michael, 58
Shipman, Pat, 4, 6, 46
Shooting in the Wild (Palmer), 106
Significance of Children and Animals, The (Myers), 122
Silent Spring (Carson), 29
skinks, 29, 31
"sky glow," 92
slacktivism, 37–40
Social Emotions Group (Northeastern University), 62
socialization, 128
social justice, 7–8, 135–37
social skills, 128
solastalgia, 35
Soulé, Michael, 8–9, 26, 36–37
South Africa, 98
species extinction
  birth control and, 26
  compassionate conservation and, 79–83
  human impact of, 135–36
  human overpopulation as cause of, 24
  rate of, 29–30
  Three Rs approach to, 81–82
  See also biodiversity loss
speciesism, 38, 47, 55
Specter, Michael, 36
spiders, 31
Spinka, Marek, 129
spiritual ecology, 55
  See also deep ethology
Spiritual Ecology (Sponsel), 61
spirituality
  animals and, 49
  indigenous, 46
  rewilding and, 54–57

sponges, 31
Sponsel, Leslie, 55, 61
starfish, 28
Steiner, Gary, 47
Sterba, Jim, 93
stereotypes, 106
St. Louis Zoo (MO), 113
Stolzenburg, Will, 29
storytelling, 107–8
stress, dealing with, 142–45
subjectivity, 4–5, 59
suburban development, 34
Super Bowl (2012), 116
survivalism, 38
survival skills, 129
Suzuki, Severn, 119–20
Sweden, 78
Switch, The (film; 2010), 107–8

Tai (elephant), 108
tailorbird, Cambodian, 31
Take Back Conservation (Foreman), 58
tamarins, golden-lion, 81
tarantulas, 31
Taylor, Paul, 52
Tedeschi, Philip, 51
television, 122–23
Ten Trusts, The (Bekoff and Goodall), 3, 27
terns, California least, 147
The Economics of Ecosystems and Biodiversity (TEEB) study, 76–77
This Is Hope (Anderson), 145–46
Three Rs approach, 81–82
Tilikum (orca), 111
tipping point, global, 146–47
Tobias, Michael, 55
Toronto (Canada), 91, 92
trauma, secondary/vicarious, 143, 144
Trauma Stewardship (Lipsky), 144
Treves, Adrian, 90
Trungpa, Chögyam, 1
Tucker, Linda, 98
Turere, Richard, 89
turtles, painted, 28

Uganda, 31, 84
United Nations, 27, 119–20
University of Colorado, 91
University of Cumbria, 127
University of Denver, 51, 136
University of Gloucestershire, 124
University of Groningen (Netherlands), 77
University of Hawaii, 55
University of Nebraska–Lincoln, 124
University of North Carolina, 56
University of Prince Edward Island (Canada), 55
University of the South, 59
University of Virginia, 91
unwilding
  alienation from nature and, 33–35
  children and, 120–21, 122–26
  defined, 33

denial and, 33, 35–37
negative effects of, 61, 132
slacktivism and, 37–40
use of term, 124
urban landscapes, rewilding of, 90–94
US Fish and Wildlife Service, 107

values, sacrifice of, 82–83
Vancouver (Canada), 91
vegetarianism/veganism, 97–98, 105
video games, 123
Vining, Joanne, 131
Virunga Volcanoes and Bwindi Impenetrable National Park (Uganda), 84

Walker, Alice, 139
Wang, Howard, 95–96
Washington, Haydn, 5
water, right to, 94
Water for Elephants (film; 2011), 108
Where the Wild Things Were (Stolzenburg), 29
White, Lynn, 57
Why the Wild Things Are (Melson), 122
wild justice, 128
wildlife, reintroductions of, 10, 81–82
wildlife biologists, 114–15
wildlife tourism, 83–85
wild play, 126
wild/wilderness
defining, 9–10, 14, 22
the outdoors vs., 57
world park preserving, 22–23
See also rewilding; unwilding
Williams, Al, 124
Williams, Terry Tempest, 8
Wilson, Edward O., 21, 34, 42, 80
Wolfer (Niemeyer), 114
wolves
controversies over, 87–90
killing of, as "self-defense," 107, 115
recovery of, 147
reintroductions of, 80, 81, 87
sensationalistic media portrayals of, 92–93, 106–7
as sentient animals, 88
Wong, Bob, 10
World of Wounds, A (Ehrlich), 5
"world park" suggestion, 22–23
Wuerthner, George, 23
Wyoming, 87

Yellowstone National Park, 81, 83
Yellowstone to Yukon Conservation Initiative (Y2Y Project), 9, 87–89

Zammit-Lucia, Joe, 58, 125
zoos, 34, 39, 52
Zoque people, 32–33

About the Author

Marc Bekoff is professor emeritus of ecology and evolutionary biology at the University of Colorado, Boulder, and a former Guggenheim Fellow. In 2000, he was awarded the Exemplar Award from the Animal Behavior Society for major long-term contributions to the field of animal behavior. In 2005, Marc was presented with the Bank One Faculty Community Service Award for the work he has done with children, senior citizens, and prisoners as part of Jane Goodall's Roots & Shoots program, and in 2009, he was presented with the Saint Francis of Assisi Award by the Auckland (New Zealand) SPCA. Marc has published around 1,000 scientific and popular essays and twenty-six books, including *Nature's Life Lessons: Everyday Truths from Nature* (with Jim Carrier), *Minding Animals*, *The Ten Trusts* (with Jane Goodall), *The Emotional Lives of Animals*, *Animals Matter*, *Animals at Play: Rules of the Game*,

Wild Justice: The Moral Lives of Animals (with Jessica Pierce), *The Animal Manifesto: Six Reasons for Expanding Our Compassion Footprint*, *Ignoring Nature No More: The Case for Compassionate Conservation*, *Jasper's Story: Saving Moon Bears* (with Jill Robinson), *Why Dogs Hump and Bees Get Depressed: The Fascinating Science of Animal Intelligence, Emotions, Friendship, and Conservation*, and two editions of the *Encyclopedia of Animal Rights and Animal Welfare*, the *Encyclopedia of Animal Behavior*, and the *Encyclopedia of Human-Animal Relationships*. He writes regular columns for *Psychology Today* (http://www.psychologytoday.com/blog/animal-emotions). His homepage is marcbekoff.com, and with Jane Goodall, http://www.ethologicalethics.org.

Phone: 415-884-2100 or 800-972-6657
Catalog requests: Ext. 10 | Orders: Ext. 52 | Fax: 415-884-2199
escort@newworldlibrary.com

14 Pamaron Way, Novato, CA 94949